The Exchange of Truth

The Exchange of Truth

Liberating the World from the Lie of Evolution

David E. Shormann, PhD

iUniverse, Inc.
New York Lincoln Shanghai

The Exchange of Truth
Liberating the World from the Lie of Evolution

iUniverse books may be ordered through booksellers or by contacting:

iUniverse
2021 Pine Lake Road, Suite 100
Lincoln, NE 68512
www.iuniverse.com
1-800-Authors (1-800-288-4677)

For a free study guide and other information,
please visit www.exchangeoftruth.com.

ISBN-13: 978-0-595-42177-0 (pbk)
ISBN-13: 978-0-595-86515-4 (ebk)
ISBN-10: 0-595-42177-6 (pbk)
ISBN-10: 0-595-86515-1 (ebk)

Printed in the United States of America

To my family, I love you.

CONTENTS

ACKNOWLEDGEMENTS

A hearty "Thank you!" to the students in my science classes and to the members of St. David's Church, Hockley, TX, for the many wonderful thoughts and comments that helped shape *The Exchange of Truth*. And a very special thanks to my wonderful wife, Karen, without whom this work would have never seen the light of day.

INTRODUCTION

*For the wrath of God is revealed from heaven against all ungodliness and unrighteousness of men, who suppress the truth in unrighteousness, because what may be known of God is manifest in them, for God has shown it to them. For since the creation of the world, His invisible attributes are clearly seen, being understood by the things that are made, even His eternal power and Godhead, so that they are without excuse, because, although they knew God, they did not glorify Him as God, nor were thankful, but became futile in their thoughts, and their foolish hearts were darkened. Professing to be wise, they became fools, and changed the glory of the incorruptible God into an image made like corruptible man-and birds and four-footed animals and creeping things. Therefore, God also gave them up to uncleanness, in the lusts of their hearts, to dishonor their bodies among themselves, **who exchanged the truth of God for the lie**, and worshiped and served the creature rather than the Creator, who is blessed forever. Amen.*

Romans 1:18-25.

"What is truth?" asked Pilate as he stood face-to-face with Christ, hours before He died for the sins of you and I. Pilate did not wait for an answer, but instead turned to address the crowd of Jews, and the rest is His story. Pilate did ask an excellent question though, one that has kept men's minds busy since the beginning.

Defining truth is simple when you have an objective standard. For example, think about a detective that specializes in finding counterfeit money. He knows a fake $20 bill when he sees it, because he has painstakingly studied a true bill down to the minutest detail. Or think about a student taking an exam. When they come to the True/False section, if they studied diligently, they will confidently breeze along. However, the less the detective and the student study the truth, the more likely they are to be ensnared by deception's sticky web.

To further understand truth, let's let one of the greatest problem solvers in history define it. The man was Leonhard Euler (1707-1783), a mathematician and passionate Christian. A father of 13 children, Euler wrote mathematics research papers at the rate of 800 pages per year[1], despite a partial vision loss starting in 1738, becoming completely blind by 1771[2]. He taught his own children and grandchildren, making scientific games for them and instructing them in Scriptures every evening[3].

In his most popular book, *Letters to a German Princess*, Euler divided truth into three classes: truths of the senses, truths of understanding, and truths of belief[4]. Truths of the senses are things we are convinced of because we have seen them, or they were revealed to us by our senses (Cardinals are red). Truths of understanding are logical things, ones we accept as true without proof (2 points define a line). Truths of belief are things we have faith in, and include all historical truths (World War 1 was a real event). While we must be careful of what we trust as truth, we should be as confident as Euler was in the one source of Truth we can always rely on, God's word revealed in Scripture.

My purpose for writing *The Exchange of Truth* was to confidently use the Truth of God's word while cautiously applying the truths of man's senses, understanding, and beliefs to uncover instances where Truth was exchanged for a lie or suppressed by evolutionary ideas.

My approach will be threefold, beginning with a study of DNA, genetics and genetic mutations. If you understand these topics, you will better visualize the unity and diversity existing in Creation. The idea of "unity amidst diversity" is foundational to the study of Creation. If we deny this attribute of God, we must deny the Trinity (Father, Son, and Holy Spirit), the Church (Body of Christ), and the covenantal relationship God has with His people.

Second, you will apply your improved biological knowledge to better understand the topics of evolution, Creation, and intelligent design. Along the way, we will take time to research the works of Charles Darwin, considered the author of evolutionary thinking. As you learn about Darwin and his predecessor Thomas Malthus, you will uncover some rarely discussed topics invaluable for destroying the false idea of evolution.

Third, you will learn how people infected by Darwinian views spread their ideas throughout society in everything from childbirth to education. Along the way, you will learn about some of history's major contributors to scientific thought, and reflect upon their views on mixing science and Scripture. You will finish by learning about the importance of renewing the mind with a Christian education.

By the end of *The Exchange of Truth*, it is my prayer that you will be more knowledgeable about God and His creation, and about why Darwinian evolution and its many forms are nothing more than bad attitudes about God and excuses for sin. As you study *The Exchange of Truth* and the works I reference, gaining knowledge and understanding, I urge you to remember that in God's eyes, if you have knowledge but have no love for your fellow man, then you have gained nothing (I Cor. 13:3). As you read and learn, constantly remind yourself that knowledge is for loving God and loving others, otherwise it is useless.

> By the end of *The Exchange of Truth*, it is my prayer that you will be more knowledgeable about God and His creation, and about why Darwinian evolution and its many forms are really nothing more than bad attitudes about God and excuses for sin.

We can destroy some enemies with bullets and large explosive devices, but we can destroy enemies of Truth with the Truth. One profound Truth we should never forget is that a man who does not love does not know God, for God is love (I John 4:8). If you read *The Exchange of Truth*, by the end you will have acquired an arsenal of mental weaponry specifically crafted for the sole purpose of wiping Darwinism from the face of the planet. Remember, though, if you attempt to conquer with a hateful heart and angry eyebrows, God will consider you as nothing.

God's kingdom is coming, and someday His will is going to be done on earth, just as it is in Heaven. However, restoring His kingdom is a work in progress, of which we are active participants, so when you finish this book, participate by purchasing this book for a Darwinist friend and discussing it over a meal. And if you have children, give them a Christian education, read Scripture and pray with them, as these are the best things you can do to help them discern Truth from fiction.

Before you begin though, you will spend two chapters learning how to think. Maybe you already think you know how to think, but I don't think you've thought of everything yet. We will focus on two areas of reasoning, deductive reasoning (applying rules) and inductive reasoning (finding rules, scientific method). So come now, let us reason together (Isaiah 1:18).

PART I
Strengthen your reasoning skills

1 THINK
Deductive Reasoning

It seems abnormal to begin a chapter with a list of definitions. However, since the famous thinker Euclid began his treatise on geometry, *The Elements,* with a list of definitions, I figured it would be okay if a mere mortal like me copied his form. Here's a list of definitions that will set the tone for the first two chapters.

Definitions

> **Philosophy:** *Phile* means love, *Sophia* means wisdom; love of wisdom
>
> **Wisdom:** knowledge, good judgment.
>
> **Knowledge:** Being aware of something.

Logic: Science of reasoning.

Logos: Godly wisdom; being aware (knowing) of God's involvement in creation, government, and redemption of the world.

Reasoning: how we think.

Think: how we form things (beliefs, opinions, procedures, etc.) in our mind.

Deductive Reasoning: Applying a rule, standard, or truth so we can know how to solve a problem, know what action to take in a situation, or discover a new rule, standard or truth. Contrast with *inductive reasoning* (Chapter 2), which is finding a rule, standard, or truth.

Syllogisms

Syllogisms are a set of statements used to create a deductive argument. Aristotle (384-322 B.C.), a.k.a. the "father of logic," is credited with developing them. Syllogisms help us understand how to think deductively, and serve as a way to determine if a rule was correctly applied, or whether an argument was valid.

Syllogisms consist of three parts: the major premise, minor premise, and conclusion. The major premise is the "rule." It identifies a property of the set. The minor premise identifies a member of the set, and the conclusion identifies the property with the member. To better understand syllogisms, study the following examples:

❑   *Valid argument and true*

    a.   *Major premise: All rainbow trout have a pink stripe.*

    b.   *Minor premise: That fish is a rainbow trout.*

    c.   *Conclusion: That fish has a pink stripe.*

In this syllogism, the *set* consists of all rainbow trout, and the *property* is their pink stripe. Put another way, our "rule" is that all rainbow trout have a pink stripe. The minor premise identifies "that fish" as a member of the set, and therefore the conclusion is "that fish has a pink stripe," which identifies the property (pink stripe) with the member (that fish). We would say that this argument is valid and true, because rainbow trout do have a pink, horizontal stripe on their side. Next is an example of a valid, yet untrue, syllogism.

❑ *Valid argument but untrue*
  a. *All frogs are green.*
  b. *Kermit is a frog.*
  c. *Kermit is green.*

Even though our argument is valid, or reasonable, it is not true because the major premise is false. All frogs are *not* green.

❑ *Invalid argument (valid if b and c switched)*
  a. *All normal horses have four legs.*
  b. *That horse has four legs.*
  c. *That horse is a normal horse*

The syllogism is invalid because the minor premise identifies a property, instead of identifying a member of the set. We cannot conclude the horse is normal just because it has four legs. It could have something else wrong with it. "Four-leggedness" is a property of normal horses, but does not automatically qualify a horse as "normal."

Syllogisms can twist your brain into a knot, and Leonard Euler knew this, so he devised some diagrams to help untangle things (Fig. 1.1). The diagrams consist of a set of 3 circles, each circle enclosing a set of items defined by the syllogism. Using the frog syllogism as an example, a large circle describes the property of "greenness." Next, since all frogs are defined as being green, a smaller circle encloses the set of "all frogs" within the green circle. Finally, since Kermit is member of the frog set, a circle for Kermit is drawn inside the frog circle, and the conclusion is made that Kermit is green. The overlapping circles of the horse syllogism signal an invalid argument.

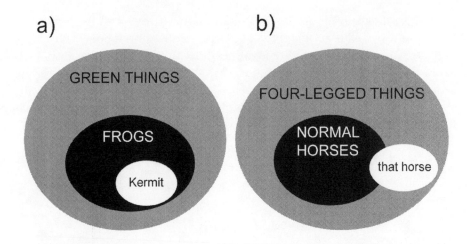

Figure 1.1. Euler diagrams of a) valid frog syllogism and b) invalid horse syllogism.

An important conclusion

Did you notice an argument can be considered valid, even if the major premise is false? Wow! That means if a major premise is assumed to be true, but is actually false, a multitude of false conclusions can blossom from it. The mind begins to race when one considers the societal impact from following one or more false major premises.

> If a major premise is assumed to be true, but is actually false, a multitude of false conclusions can blossom from it.

Syllogisms are simply a step-by-step method for finding new rules, and familiarity with them will allow us to better discern truth from fiction.

Examples of uses of deductive reasoning in our lives

The following are examples of how man uses deductive reasoning.

Geometry: Euclid lived around 300 B.C. and produced a book called *The Elements*. The book was an attempt at organizing all the Greek mathematical achievements of the previous 300 years into one volume. Starting with 23 basic definitions and 10 axioms and postulates (Fig. 1.2), or statements assumed to be true without proof, Euclid went on to deduce hundreds of geometric theorems. The axioms and postulates are like major premises of a syllogism, with theorems as the conclusions. The Greeks believed these axioms or postulates and the theorems deduced from them were true because human reasoning alone proved them to be true. They did not believe all knowledge came from God (Proverbs 2:6). Although their geometric developments were profound, the Greeks failed to incorporate key concepts like infinity[1]. Greek mathematics and science stagnated because their theological ponderings were not biblical[2].

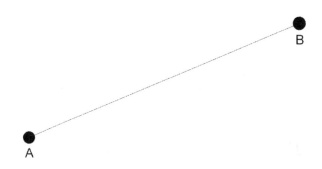

Figure 1.2. Euclid's 1st Postulate: two points determine a unique straight line.

Measurement: "Differing weights and differing measures, the Lord detests them both" (Proverbs 20:10).

God detests dishonest standards of measurement, or in other words, dishonest application of deductive reasoning. Many folks believe science and religion are opposing views, but if mathematics is the language of science, and mathematics includes measuring things, and God detests dishonest measurements, then how can we say we can't mix science and religion? Secular biology textbooks make it clear that we should separate the natural and the supernatural. For example, on p. 15 of *Biology, the Unity and Diversity of Life*, a footnote at the bottom of the page quotes "systematic observations, hypotheses, predictions, tests. In all these ways, science differs from systems of belief that are based on faith, force, or simple consensus."[3] This is a common "exchange of truth" found in secular biology textbooks.

Proverbs 20:10 is a clear example of the supernatural guiding the natural. Therefore, we should look for historical examples where people and nations applied Proverbs 20:10, because measurements are one of the most prevalent forms of deductive reasoning in our world.

Let's begin with Queen Elizabeth (1533-1603). When she became Queen in 1558, she quoted Psalm 118 "This is the Lord's doing and it is marvelous in our eyes." She felt chosen to reign by God's will[4], and one of her priorities was establishing a set of standard weights and measures. The critical unit of measurement was the yard, and in 1601 a brass yardstick was constructed as a national standard. Elizabeth ordered new standards made for weights and measures. She sent copies of the standards to fifty-eight market towns with instructions that a description of them be pinned up in every church and read during the service twice a year for the next four years[5].

Precise measurement was critical to those colonizing America, where accurate land surveys were a high priority. Edmund Gunter, a contemporary of Elizabeth, developed what is now known as Gunter's chain. The chain contained 100 links and was exactly 22 yards long. The beauty of Gunter's chain was in its ease of use for measuring area, because an area of 10 chains by 1 chain equaled exactly one acre. Therefore, to determine acres, one simply needed to divide the number of square chains of area by 10.[6]

Since "The Lord abhors dishonest scales, but accurate weights are his delight," then an attribute of God is an accurate system of measurement. Applying standards in the form of accurate measurement is an attribute of God, and science is nothing without measurement. Therefore, to say science and religion cannot mix is to exchange truth for a lie.

> To say science and religion cannot mix is to exchange truth for a lie.

Physics: Isaac Newton (1642-1727) used Euclid's deductive system found in *The Elements* to write *The Principia*. Like Euclid's work, his volume begins with basic definitions and then states three "laws of motion." Newton was a master of the deductive reasoning process, believing science and religion were completely compatible. A brief read-through of *The Principia*, arguably the single most important contribution to scientific knowledge, will reveal quotes like the following: "This most beautiful system of the sun, planets and comets, could only proceed from the counsel and dominion of an intelligent and powerful being"[7]. Newton's laws help us understand everything from how fast an apple falls to the ground to how a jet engine can propel an aircraft through the sky.

Analogies: *Analogous* means similar, and therefore analogies are statements containing some abstract (general) truth which help us discover the truth in a similar, or analogous, statement. Analogies are a form of deductive reasoning, because you apply the abstract truth found within them in a new situation. Jesus' parables are great examples of analogies, and Luke 8:4-15 is one that we will use in Chapter 3.

Why do we think deductively?

Why has deductive reasoning been around so long? Because that's how God made us! Man's first use of deductive reasoning is Genesis 1:28, thousands of years before Aristotle and Euclid. In Genesis 1:28, God commands us to rule over His creation. In other words, one of God's first rules for us was to rule. God's word contains a treasure trove of major premises we can confidently apply to help us discern truth from fiction. His word guides us and helps us answer the question "How should we live?" (2 Timothy 3:16-17). God's word should be the most important major premise in our lives, because the conclusions we draw from Scripture will lead us down the right path.

An appropriate major premise for the study of science

Francis Bacon (1561-1626) played an important role in developing the scientific method[8]. His father was the second-most important counselor to Queen Elizabeth. Exceptionally bright, Bacon received a home education before entering Trinity College Cambridge at the tender age of 12. At 33, for fun during Christmas vacation he wrote down all the quotes he had memorized, totaling over 1,600 items[9]. In *The Advancement of Learning, Book 1,* Bacon stated the following as a major premise for learning:

> *A man cannot be too well studied in the book of God's word or in the book of God's works, divinity or philosophy, but rather endeavor an endless progress or proficience in both; only let men beware that they apply both to charity, and not to swelling, to use and not to ostentation; and again, that they do not unwisely mingle or confound these learnings together.*[10]

The passage will serve as one of the most important major premises in my book. Take some time to memorize it. Bacon's words are analogous to Jesus' words in Matthew 22:29:

> *You are in error because you do not know the Scriptures or the power of God*

Scripture is God's revealed word, and *the power of God* is His created works. Bacon is often criticized for his overemphasis on experimentation and observation, and "Baconianism" has manifested itself in many ways, including the hostility some modern churches show towards creeds and confessions, while simultaneously overemphasizing the worship "experience"[11].

Bacon's overemphasis on experimentation was probably a response to what he perceived as an overemphasis on deductive reasoning in the universities. Bacon believed their methods resulted in minds being "shut up in the cells of a few authors (chiefly Aristotle their dictator)"[12], as the words of these authors were rarely challenged[13].

While Bacon had some faulty major premises, he also had some exceptional ones. Just like it is unreasonable to ignore all of David's Psalms because of his many sins, we shouldn't ignore all of Bacon's words just because he developed some faulty major premises. When you read Bacon's works, it is obvious he was, like David, a man after God's own heart.

Eliminating erroneous thinking and improper major premises regarding nature will only occur when men engage in a unified study of His word and His works. Christians must stop being confused by viewpoints proclaiming science and religion (specifically Christianity) don't mix, and should boldly proclaim as Bacon did that studying His word and His works together is appropriate.

An inappropriate major premise for the study of science

Charles Darwin (1809-1872), is most famous for his theory of natural selection. He was an observant man, well versed in the scientific method (He said "science consists in grouping of facts so that general laws or conclusions can be drawn from them").[14] However, he lacked many attributes common to good scientists, particularly naturalists. Dissection disgusted him and he was a poor artist[15]. Surprisingly, he was poor at mathematics[16]. Most of the last 40 years of his life, he was an invalid, and while the reasons behind this hermit-like existence are not fully understood, evidence suggests a psychological problem of some sort[17].

An amusing anecdote in his life is that he almost did not participate on the historic voyage of the H.M.S. Beagle because of the shape of his nose. It seems that Captain Robert Fitz-Roy, convinced he could judge a man's character by the outline of his features, doubted whether anyone with Darwin's nose "could possess sufficient energy and determination for the voyage."[18] Needless to say, Fitz-Roy overcame his strange beliefs and allowed Darwin to come aboard as the ship's naturalist.

At one time Darwin had studied for the clergy, and did not "in the least doubt the strict and literal truth of every word in the Bible"[19]. But then,

Disbelief crept over me at a very slow rate, but was at last complete. The rate was so slow that I felt no distress, and have never since doubted even for a single second that my conclusion was correct. I can indeed hardly see how anyone ought wish Christianity to be true[20].

And this is the major premise, the worldview that spawned the theory of natural selection, a theory whose major purpose was "to show that species had not been separately created" and to "overthrow the dogma of separate creations"[21].

In a nutshell, Darwin believed there was no unity between the study of science and the study of religion. This is in direct contrast to Francis Bacon's major premise, and more importantly, in direct contrast to Scripture. Unfortunately, secular biology textbooks acknowledge Darwin's theory of natural selection as science's "unifying principle." One biology textbook I will refer to on occasion is aptly named *Biology: The Unity and Diversity of Life* by Cecie Starr and Ralph Taggart. High school AP Biology classes as well as college introductory biology classes use the text, and, minus the evolutionary dogma, it is beautiful, well-written and well-researched. The "unifying concept" referred to in the title is revealed on p. 3:

Unity underlies the world of life, for all organisms are alike in key respects.... Theories of evolution, especially the theory of evolution by natural selection as formulated by Charles Darwin, help explain life's diversity. The theories unite all fields of biological inquiry into a single, coherent whole.[22]

According to Randall Hedtke, former public school science teacher and author of *The Great Evolution Curriculum Hoax*, using evolution as the "unifying theme" in biology textbooks has been standard since the early 1960's[23]. Starr and Taggart's view reflects the norm among today's biology textbooks. Evolutionary ideas do not stop at biology classroom doors, however, but rather spill out and leak into practically every course and degree plan offered at secularized institutions.

Acknowledging Darwinism as the reason for the unity and diversity surrounding us is a huge exchange of truth, and therefore an inappropriate major premise for the study of science. Did you notice how Darwin described his journey away from Biblical truth? Disbelief slowly crept over him, until he came to the same point Eve did in the Garden, answering the question "Did

God really say …?" with a resounding "NO!" that ever since has sent many wise men down a foolish path of deception.

Chapter 1 Summary

Many facts contained within this book are currently illegal to teach in most public schools, but they are facts everyone has a right to know. If you are in high school, the major premises shaping your worldview today will have a profound affect on the rest of your life and on the people you know. These premises will serve as filters to help you make logical deductions.

If you don't think your choice of major premises matters, consider this amazing statistic: Of students who enter college proclaiming they are born-again believers, 50 percent say they are no longer so upon graduation[24]. How can this be? I think the best answer to this question lies in the separation of Christianity from education, and its replacement with Darwinism.

Late in the 12th century, a phenomenon unique to Europe materialized, *the university*. Originally, these schools owned no real estate, but were instead an association of teachers or students. Although not always theologically or scholarly accurate, what under girded the university was the idea that an all-encompassing worldview united all subjects. Christianity provided unity for all departments of study[25].

In the 21st century, most American universities, like most biology textbooks, have exchanged the unity of a Christian worldview with an evolutionary worldview. The major premises taught currently are either atheistic (deny existence of God) or agnostic (cannot know if there is a God), and according to the above statistics, universities are effective at getting students to believe their major premises.

How can you avoid the confused teachings prevalent in today's educational system? Psalm 119:11 says "I have hidden your word in my heart that I might not sin against you." Scripture memory is one of the most effective methods for avoiding confused thinking and false major premises. Follow the educational example of Francis Bacon and his Christian humanist contemporaries, who not only memorized Bible passages, but memorized worthwhile

> Scripture memory is one of the most effective methods for avoiding confused thinking and false major premises.

quotations of other men. As you study His word and His works together, you will learn that a Christian view of a rational and personal world was the basis for the birth of science, and is the only real foundation for pursuing science in our day[26].

www.snowflakebentley.com

2 THINK HARDER
Inductive Reasoning and Critical Thinking

Inductive Reasoning

In Chapter 1, we defined deductive reasoning as *applying rules*. Contrast that with inductive reasoning, defined as *finding rules*. Interestingly, the history of the world begins with a command for man to use both deductive and inductive reasoning. God tells us in Genesis 1:28 to manage His creation, but doing so requires a management plan. How should we go about finding rules to help us manage His creation?

More than anything, observing God's works will help us find the best rules for managing His creation. Observation of God's works is often referred to as *science*. Science comes from the Latin *scientia* meaning having knowledge. When applied to second causes, *science* means knowledge gained through observing the physical

> Prior to Darwin, "second causes" was a common phrase describing God's works and the laws He created to govern Creation. God was considered the "first cause."

world, that is, God's works. Darwin gave a good explanation for science when he said "Science consists in grouping of facts so that general laws or conclusions may be drawn from them"[1].

The Scientific Method

Just like a syllogism is a step-by-step method for applying a rule, the *scientific method* is a step-by-step method for finding a rule. Francis Bacon (1561-1626) was one of the founders of the scientific method. Bacon thought the universities of his day deceived men with hollow and deceptive philosophies (Colossians 2:8), mistaking for truth what is falsely called knowledge (I Timothy 6:20). During Bacon's day, and again in ours, universities commonly taught that deductive reasoning based on man's reasoning alone was the primary pathway for gaining knowledge. Bacon believed knowledge arrived at this way, while good at first, would never rise higher than the man who developed it[2]. He believed reading books alone would not get us far, which Solomon, the wisest man to walk the earth, believed as well (Eccl. 12:12). As his foundation for the scientific method, Bacon looked to the mechanical arts, where "the first deviser comes shortest, and time addeth and perfecteth"[3].

Today, we define the scientific method as *a systematic (step-by step) way to think inductively (find rules)*. It is simply a way to answer a question by performing an experiment. The scientific method begins and ends with a question, resulting in a continual advancement of learning. Here are its 5 steps:

1. **Question:** After observing something, whether a meteor, a manatee or a magnet, questions arise in our minds about what we just observed. Every science experiment begins with a question. To carry out God's command in Genesis 1:28, man must ask questions. The sole purpose of the scientific method is to provide an outline for answering a question.

2. **Hypothesis:** The simplest definition for a hypothesis is "educated guess." When we develop a hypothesis, we are guessing

at the answer to our question. This is not a blind guess, this guess requires one to read, research and discuss the question with experts beforehand. The major premises or worldview used to generate knowledge greatly affects the hypothesis. It is an important part of the scientific method, because if the major premises used to develop the hypothesis were false, then the scientific conclusions will be false.

The hypothesis is where deductive and inductive reasoning merge. The hypothesis is similar to the major premise of a syllogism, except we do not assume the hypothesis to be true without proof. We test the hypothesis using the scientific method, and what we believe about the results of our experiment depends on how much we trust the information used to educate ourselves and develop our worldview. Darwin said "it is a fatal fault to reason whilst observing, though so necessary beforehand and so useful afterwards"[4]. This is excellent advice for any scientist.

When conducting an experiment, we should focus on collecting the data, eliminating bias when possible. We should never cheat and manipulate the data so it fits our worldview. However, keep in mind the reason any scientist conducts an experiment is to try to prove a hypothesis that was influenced by their worldview! Being well versed in the book of God's word and God's works will help you build your hypotheses on a solid-rock foundation rather than a shifty, sandy one based solely on the words of men.

3. **Methods:** This section consists of a step-by step experimental procedure, written with enough detail that someone else could repeat your experiment. It is like writing a recipe, where your goal is to describe precisely enough that someone can make it the same way you did. The *Methods* also contain a listing of materials used to perform the experiment.

4. **Results:** When performing an experiment, scientists observe and record information, or *data*, about it, which is later analyzed and processed. Scientists display the *Results* of the experiment in a useful format such as a table or graph, and present it in a way that individuals who know nothing about the experiment could observe it and draw a proper conclusion.

5. **Discussion:** As the last section of the scientific method, the scientist discusses whether the hypothesis was correct and why.

Also, he lists any sources of error, such as instrument error, measurement errors, etc. Good science experiments result in more questions to answer using the scientific method. The *Discussion* section is the place to present those questions.

The goal of the scientific method is the same as the goal of its founder, Francis Bacon, who wished to see a continual advancement of learning. We cannot know everything that God knows, and much will always remain a mystery, but if we humbly seek Him, he will answer our questions and teach us great and unsearchable things that we do not know (Jeremiah 33:3).

Does God want us to think inductively?

Yes! One of God's first commands to man in Genesis 1:28 was to think scientifically! He wants us to find rules for properly managing His creation. However, since man is imperfect, we must be careful when finding second causes, or the rules we come up with could develop into hollow and deceptive philosophies (Colossians 2:8). God is the perfect First cause, and only by His grace can we know anything about the second causes.

Francis Bacon hits the nail on the head again when, after discussing Matthew 22:29, he says that to remove error we must "study first the Scriptures, revealing the will of God, and then the creatures expressing his power"[5]. Notice the order, Bible study first *then* creature study. Bacon suggested that Scripture, and in particular the Psalms "do often invite us to consider and magnify the great and wonderful works of God"[6].

Being "well versed" in a few dozen Psalms is a good foundation for inductive thinking. Also, good scientists are observant, and reflection on God's word will lead to proper conclusions regarding observation of His works.

> The Psalms do often invite us to consider and magnify the great and wonderful works of God.
> Francis Bacon

If Bacon's assessment of learning is correct, then examples of exceptional scientists who were students of the Bible as well as students of Creation should abound. Indeed a multitude of examples exist, men like Copernicus, Kepler, Galileo, Kelvin, Faraday, and Boyle, to name a few. We have already discussed Leonhard Euler, considered by other scientists and mathematicians of his time as "the master of us all"[7], and Isaac Newton, who contributed perhaps the single-most important work in the history of science. Now let's learn about a few not-so-well-known scientists, their methods, and their incredible contributions.

Charles Goodyear (1800-1860)[8,9]: From an early age, Charles was intensely religious and a self-assigned student of the Bible. His eldest daughter, Ellen, always remembered him fondly and recalled that he regularly read passages of the Bible to the children. During Goodyear's time, rubber was just coming on the scene, but it was leaving as quickly as it came because it melted in the summer heat and grew hard and brittle in the winter cold. When his hardware business collapsed, Charles made his mark on the world by inventing a process that would make rubber more durable, known as vulcanization.

Charles arrived at the process mainly through trial and error, meaning he used many "uneducated guesses" since so little was known about rubber. He believed that God had chosen him to solve the great rubber mystery, and in his own words he was "the instrument in the hands of his Maker." How did Goodyear know that he was the man to solve this problem of rubber's instability? I think it was because of his unwavering faith in God.

Today, the Goodyear Tire & Rubber Company, named in his honor, is the world's largest rubber business. Rubber is now an unbelievably important resource, with one cultivated rubber tree existing for every two people on earth. With the invention of automobiles came a need for rubber tires, hoses, seals, etc. Without Goodyear's invention, spreading the Gospel message would be much more difficult indeed (Fig. 2.1).

Figure 2.1. Good tires are helpful!

Matthew Maury (1806-1873)[10,11]. Matthew Maury's grandfather, the Reverend James Maury, was an Episcopal clergyman, and instructor of youth at a school where Thomas Jefferson began his education. Matthew's father, Richard Maury, would assemble the children every morning and evening to read the Psalm for the day. Matthew became so familiar with the Psalms that later in life he could quote them, chapter and verse. Inscribed on his tombstone at the U.S. Naval Academy is Psalm 8:8: "… all that swim the paths of the seas." Maury believed that since God said there were paths in the sea, he should be able to find them.

As a Christian, for Maury the paths in the sea were more than an "educated guess," they were a truth to discover. In 1855 he wrote the *Physical Geography of the Sea*, which many consider as the first oceanography textbook. Many honor Maury as the "Father of oceanography." Maury's major premise was that Scripture was true and God was the Creator, and this affected his data collecting and his conclusions about that data. It was natural for him to say things like "whether of the land or the sea, the inhabitants are all His creatures, subjects of His laws, and agents in His economy"[12].

> Whether of the land or the sea, the inhabitants are all His creatures, subjects of His laws, and agents in His economy
>
> Matthew Maury

While a hard worker described as the "indefatigable investigator" (by 1854 he had collected data on 380,284 observations), Maury was also a family man, and especially liked young people. One of his daughters recalled he never had a private study, preferring to work amid his wife and children wherever they congregated. Maury applied his learning to "use and not to ostentation"[13] and provided a great service to others by making sea travel safer and more efficient.

Wilson Bentley (1865-1931)[14]. From early childhood, Wilson Bentley was obsessed with snow. A major premise of his was Job 38:22 "Have you entered the treasury of snow?" On his 17th birthday, his parents gave him a special camera-microscope, which he used to photograph snowflakes for the next 49 years. Bentley kept meticulous records of his work, so detailed in fact, that we know he took exactly 5,381 pictures of snow crystals (snowflakes are actually clumps of snow crystals), no two of which were alike. Bentley was confident that God had placed him in Jericho, Vermont, one of the world's best places to study snow, to carry out His purposes. In a magazine article, Bentley wrote "so all but infinite is the number of individual crystals produced by a single snowstorm, and so limitless the variety of form and structure of these gems from God's own laboratory." Bentley showed the world that snowflakes are yet another example of God's attribute of "unity amidst diversity." All six-sided, all made of water, yet no two alike.

Inductive Reasoning summary:

Inductive reasoning is about finding rules, and the scientific method is a systematic approach to inductive reasoning. Deductive and inductive reasoning merge in the *hypothesis*, and our major premises have a great effect on the hypotheses we develop. Charles Darwin understood science and the scientific method, but his false major premises (God did not create separately, don't know if God exists) created many errors in his scientific conclusions, which we will discuss later.

> Charles Darwin understood science and the scientific method, but his false major premises created many errors in his scientific conclusions.

Christian scientists like the ones we just studied differed from Darwin in their major premises, and they were confident their work was God's will for their lives, laboring in service to Him and their fellow man. As a result, their contributions were quite beneficial to mankind. We should be as confident as they were that God has a plan for each of us, and we should be just as committed to carrying out that plan. A major premise we should faithfully proclaim is "with God, all things are possible" (Matthew 19:26).

Critical thinking:

The "critical" in critical thinking does not mean "crucial," or "high priority," it means "criticize." *To think critically means to question.* For example, if someone told you "I have a pet frog that is purple and spits fire," if you were thinking critically, you may ask questions like "Where did you get your frog?", "What kind of frog is it?" "Where is it from?", "Is this a real frog?", "May I see your frog?" The Web site, www.criticalthinking.org, says "thinking, to be critical, must not be accepted at face value but must be analyzed for its clarity, accuracy, precision, relevance, depth, breadth, and logicalness." The Web site describes these statements as "universal intellectual standards," which should "help guide students to better reasoning."

Since the essence of critical thinking deals with questioning, when we hear or read something, we should ask questions like "Could you elaborate further?", "How do you know what you are saying is true?", "Could you be more specific?", "How are your statements connected to the question?" "How are you considering the problems with your question?" "Do we need to consider another point of view?" "Does this make sense?" Since scientific thinking also starts with questioning, it is a form of critical thinking. Scientists test hypotheses by experimentation, and use the results to draw conclusions. Critical thinking, like scientific thinking, is a systematic method of inquiry.

History of Critical thinking

While not given credit in the "history of the idea of critical thinking" section on www.criticalthinking.org, obviously *Solomon*, who lived around 950 B.C. and wrote most of the book of Proverbs, understood critical thinking. For example, Proverbs 14:8 describes the essence of critical thinking: "The wisdom of the prudent is to understand his way." To understand if you are going the right way, you have to think about it and ask questions like "is this the right way?", "how can I be sure?"

Socrates lived around 450 B.C., and questioned common beliefs and explanations of his day. He is the man most scholars consider as setting the agenda for the tradition of critical thinking. Socrates believed "the unexamined life is not worth living" and "no man knowingly does evil."[15]. He believed correct action involved thought, so someone who does evil did not know they were doing evil. They put Socrates on trial, charged him with corrupting the young and showing disrespect for religious traditions, and sentenced him to death by poisoning.

Jesus Christ lived around 0 B.C. While no one in secular critical thinking circles typically gives Him credit for developing critical thinking, Jesus definitely "put thought to his ways," and questioned even the motives of God (Matthew 27: 46). Jesus died on the cross, sentenced to death because He questioned the religious traditions of the time.

Another important figure in the history of critical thinking was *Francis Bacon* (1561-1626). www.criticalthinking.org says that his book, *The Advancement of Learning*, "is considered one of the earliest texts in critical thinking, for his agenda was very much the traditional agenda of critical thinking." We have already researched Bacon's agenda in some detail, and we know he emphasized studying "first the Scriptures, revealing the will of God, and then the creatures expressing his power"[16]. In Bacon's mind, the continual improvement in our knowledge would only happen by unifying and putting into action our knowledge of Scripture and of Nature. Bacon despised those who thought human reasoning alone was the correct way to think. This was a prevalent form of teaching during his time, which he considered a "disease of learning." He was certain that learning could only advance by reading a variety of books including the Bible, as well as contemplating God's works. He considered the "undertakers of learning" to be people whose

pride inclined them to leave the oracle of God's word and to van-
ish in the mixture of their own inventions, so in the inquisition
of nature they ever left the oracle of God's works and adored the
deceiving and deformed images which the uneven mirror of their
own minds or a few received authors did represent unto them[17].

One of Bacon's most important major premises was that human reasoning alone would not advance learning, and a unified study of "divinity and philosophy" was the only way to advance learning.

Critical thinking today

Current teachings of critical thinking involve human reasoning alone as the source of knowledge, which is the method Bacon condemned. Therefore, there is an "exchange of truth" on the criticalthinking.org Web site, because Bacon's agenda was not the agenda of today's form of critical thinking. This is obvious because the one question not asked on this Web site is "What does Scripture say about this?", a question Bacon would have asked when thinking critically. The 1998 article titled *Critical Thinking: What It Is and Why It Counts*, defines a good critical thinker as someone who is "trustful of reason, open-minded"[18], among other things. Unfortunately, this article did not list "trustful of Scripture" as one of the attributes of a good critical thinker.

Another huge exchange of truth in today's version of critical thinking is the lack of emphasis on memorization. In the report *Critical Thinking, What It Is and Why It Counts*, when trying to define critical thinking the author says " you don't really want a definition plopped on the page for you to memorize, do you? That would be silly, almost counterproductive"[19]. According to this report, proper thinking emphasizes mathematical reasoning, historical reasoning, geographical reasoning, etc. I doubt that Francis Bacon, who wrote down over 1600 memorized quotations just for fun, would agree at all with today's system of critical thinking!

Time to get back to a Biblical view of critical thinking

Jesus rebuked those who did not know God's word or His works (Matthew 22:29), as well as those who did not know God's word but did know God's works: "Hypocrites! You know how to discern the face of the sky, but you cannot discern the signs of the times" (Matthew 16:2).

> Our primary purpose for acquiring knowledge should be to glorify God and to relieve man's condition.

What should be our primary use of knowledge? According to Bacon, our primary purpose for acquiring knowledge is to glorify God and to relieve man's condition, making life better for mankind. Secondary uses of knowledge include satisfying our curiosity, winning an argument, or earning a living[20]. There should be a balance to our learning, not overly relying on man's mind[21], and not being excessively critical (like Socrates), or not critical enough[22]. Following God's word and praying for His will to be done helps us discern when to stop and start the questioning. If we don't know His will, we can never stop questioning, and our lives will not advance but will stagnate in a quagmire of unanswered questions.

Chapter 2 summary:

Critical Thinking is scientific thinking, because both are about asking questions. The scientific method provides a framework for thinking inductively, just like syllogisms provide a framework for thinking deductively. Critical thinking is a great concept, but the modern definition is flawed because it is based on the false major premises that human reasoning is superior, and that memorization is not necessary. Real critical thinking involves having a firm foundation based on the study of God's word and God's works, God's works including man and the study of man, or humanities. We should memorize as much as possible, or at least know where to find the information we need to answer our questions.

They persecuted Socrates when he questioned the religious leaders of his day. They persecuted Jesus when he questioned the religious leaders of his day. They persecuted Galileo when he questioned the religious leaders of his day (the Catholic church, who accepted a geocentric view of the universe based not on Scripture, but Aristotle, which differed from the observations made by Galileo)[23]. They could persecute *you* for questioning today's leaders who have exchanged the truth of God for a lie.

If the Author of truth could not convince those around him that He was who He said He was, you will have an even harder time convincing everyone you meet that God is the Creator, the Author of Science, and the Author and perfector of our faith. God gives humans freedom to choose. He even gave this freedom to angels, as we know Satan rebelled against God.

Therefore, because God has given us such liberty, we should expect opposition and setbacks, but we can also be confident of many victories. In the words of Leonhard Euler "God.... places men, every instant, in circumstances the most favorable, and from which, they may derive motives the most powerful, to produce their conversion"[24]. Pray that God will use you and that He will put the words in your mouth that may lead someone to exchange a lie for truth, and if necessary, lead them to salvation. If it is you that needs to repent of your sins, then don't wait a minute longer. If you repent and confess with your mouth that Jesus is Lord, He will save you from eternal death (Romans 10:9). Remember also the stories of the scientists we have studied, and be confident that studying His word together with His works is the right way to learn. Trust that He has a special plan for your life which you won't reveal by reasoning alone.

PART II
Biology basics

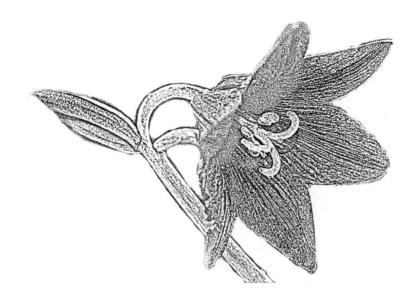

3 MOLECULES AND MENDEL
Fundamentals of DNA and Genetics

In chapters 1 and 2 you learned some reasoning skills foundational to under-standing science and the scientific method. For the next two chapters, apply your new skills to the study of life, or biology, for that is what *biology* means, "the study of life." In Chapter 3 you will learn about DNA and genetics, and in Chapter 4 you will study gene mutations and mathematical probability. Having a good understanding of these biological and mathematical basics will allow you to discern truth from fiction in biology and beyond.

As you read the next two chapters, you can avoid much confusion by mem-orizing the definitions listed at the beginning of the sections on DNA and genetics. First though, let's begin by reading a parable of Jesus. Reflect on this parable as you read the chapter, and we will discuss its relevance to our study at the end of the chapter. Luke 8: 4-15 says:

And when a great multitude had gathered, and they had come to Him from every city, He spoke by a parable: "A sower went out to sow his seed. And as he sowed, some fell by the wayside; and it was trampled down, and the birds of the air devoured it. Some fell on rock; and as soon as it sprang up, it withered away because it lacked moisture. And some fell among thorns, and the thorns sprang up with it and choked it. But others fell on good ground, sprang up, and yielded a crop a hundredfold." When he had said these things He cried, "He who has ears to hear, let him hear!" Then His disciples asked Him, saying, "What does this parable mean?" And he said, "To you it has been given to know the mysteries of the kingdom of God, but to the rest it is given in parables, that seeing they may not see, and hearing they may not understand (Isaiah 6:9).

Now the parable is this: The seed is the word of God. Those by the wayside are the ones who hear; then the devil comes and takes away the word out of their hearts, lest they should believe and be saved. But the ones on the rock are those who, when they hear, receive the word with joy; and these have no root, who believe for a while and in time of temptation fall away. Now the ones that fell among thorns are those who, when they have heard, go out and are choked with cares, riches, and pleasures of life, and bring no fruit to maturity. But the ones that fell on the good ground are those who, having heard the word with a noble and good heart, keep it and bear fruit with patience.

*Unless noted otherwise, all discussions involving cells pertain to human cells.

DNA

Terms to memorize

a. **DNA**—Deoxyribonucleic acid. Large molecule shaped like a twisted ladder. Each "step" of the ladder contains a pair of nucleotides, joined at the middle by a hydrogen bond. One side of the ladder, called a DNA strand, contains instructions to make enzymes and other proteins needed to build and maintain cells.

b. **Nucleotide base**—consists of a phosphate molecule, a deoxyribose sugar, and an amino acid (either adenine, thymine, guanine, or cytosine).

c. **Base pair**—A pair of nucleotide bases, joined at the middle by a hydrogen bond. Makes a "step" on the DNA ladder. Adenine always pairs with thymine, and cytosine always pairs with guanine. The DNA in just one normal human cell contains 3.2 billion base pairs.

d. **RNA**—Ribonucleic acid. Similar to DNA (ribose sugar instead), and important in protein manufacture.

e. **Amino acids**—molecules that are the building blocks of proteins. Made primarily from nitrogen, carbon, oxygen, and hydrogen. There are 20 naturally occurring amino acids.

f. **Proteins**—chains of amino acids. Examples of proteins include enzymes, hair, the lens in your eye, and countless other things.

g. **Enzymes**—A protein that acts as a catalyst, which means it participates in chemical reactions and speeds them up, but it is not destroyed in the process. Some enzymes help manufacture new chemicals, while other enzymes help break down chemicals, as in digestion.

h. **Codon**—A set of three nucleotide bases on a strand of DNA that act as "code words." Since there are 4 bases in DNA and 3 "letters" in each code word, a total of $4^3 = 64$ code words exist in the DNA language. Most of the code words refer to a certain amino acid, but some signal the beginning or ending of protein manufacture.

What does DNA do?

At the moment of your conception, an egg cell, containing 23 of your mom's chromosomes, and a sperm cell containing 23 of your dad's chromosomes united and began duplicating. Incredibly, now your body contains billions of cells, all duplicated from that original cell, and inside the nucleus of every one exists a copy of the original 23 chromosome pairs.

Chromosomes are a single strand of DNA (Fig. 3.1). Perhaps you have seen a photo or picture of chromosomes where they looked like a little "x," but this shape only occurs for a short duration during cell division. When a cell is not dividing, the DNA is stretched out into long strands. When it is time for the cell to divide, one of the first actions is the DNA replicates. The DNA "unzips," and free nucleotides in the surrounding fluids come in and attach on the unzipped portion. Since adenine only pairs with thymine and cytosine only pairs with guanine, two identical copies of DNA form, one for each new cell.

Between periods of cell division, or *mitosis*, the DNA actively participates in the manufacture of proteins (Fig. 3.2). *Every* protein is synthesized in accordance with instructions in DNA[1]. During protein manufacture, only portions of DNA, called *genes*, unzip. The sequence of nucleotide bases on some genes varies little from person to person. However, enough different genetic information exists so no two peo-

> Genes contain the instructions for making you, and since half of these instructions came from mom and half from dad, you look a little bit like mom and a little bit like dad, but not exactly like either one.

ple (except identical twins) have the same DNA. Genes contain the instructions for making you, and since half of these instructions came from mom and half from dad, you look a little bit like mom and a little bit like dad, but not exactly like either one.

Protein manufacture, or synthesis, requires two steps, *transcription* and *translation*. Transcription begins at a *promoter*, a DNA base sequence signaling the start of a gene. An enzyme assists in unzipping and combining nucleotides to form a strand of messenger RNA (mRNA). The nucleotide pairing is the same for RNA as for DNA, except when adenine is the base on DNA, uracil (not thymine) pairs with it on the RNA side.

After reaching a certain point on the gene, the mRNA molecule releases, carrying the transcribed message over to a cell organelle called a *ribosome*, and translation begins. At the ribosome, mRNA combines with transfer RNA (tRNA). Each tRNA molecule has an amino acid attached to a triplet of nucleotides known as an *anticodon*. The anticodons find their matching codon on the mRNA, and as they link up, the new protein forms. The amino acids link with peptide bonds, and when translation finishes, the protein releases into the cell.

Amazing facts about DNA

Scientists define *genomes* as the DNA contained in a haploid (half) number of chromosomes for a given species. The genome of a human contains about 30,000 genes made up of 3.2 billion base pairs! Let's look at some numerical facts about just one chromosome, chromosome number 22[2].

- 48 million base pairs
- 700 genes
- Smallest gene is 1000 base pairs long
- Largest gene is 583,000 base pairs long
- Average gene is 19,000 base pairs

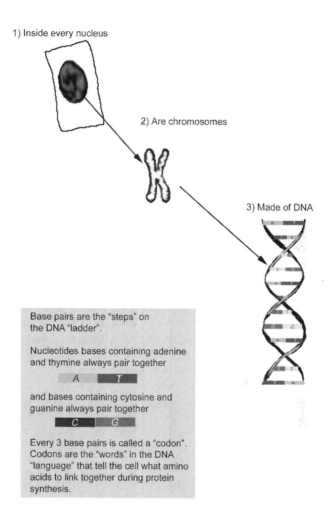

1) Inside every nucleus

2) Are chromosomes

3) Made of DNA

Base pairs are the "steps" on the DNA "ladder".

Nucleotides bases containing adenine and thymine always pair together

A   T

and bases containing cytosine and guanine always pair together

C   G

Every 3 base pairs is called a "codon". Codons are the "words" in the DNA "language" that tell the cell what amino acids to link together during protein synthesis.

Figure 3.1 Location and structure of DNA.

1) During cell division, or mitosis,

duplicated chromosomes
migrate to opposite sides

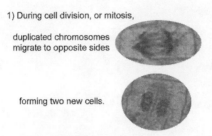

forming two new cells.

2) When the cell is not dividing, it is making proteins,
which requires two steps.

a) During *transcription*, a gene "unzips", and forms
messenger RNA (mRNA).Cytosine still pairs with guanine,
but adenine pairs with uracil on the mRNA side.

mRNA

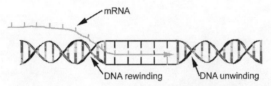

DNA rewinding   DNA unwinding

b) During *translation*, mRNA links with transfer RNA, and a
protein chain forms.

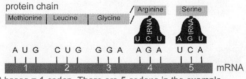

protein chain

| Methionine | Leucine | Glycine | Arginine | Serine |

tRNA   tRNA
U C U   A G U

A U G   C U G   G G A   A G A   U C A
  1   2   3   4   5   mRNA

3 bases = 1 codon. There are 5 codons in the example.

Figure 3.2 Cell processes during and after cell division

Since every three base pairs is a "code word" for a specific amino acid, chromosome 22 can make proteins ranging from about 333 to almost 200,000 amino acids long! Keep in mind we are just talking about one chromosome. Even a microscopic bacterium like E. coli has 4,279 genes in its genome. We will discuss the significance of these numerical facts when we cover mutations and probability in Chapter 4.

Why concern ourselves with DNA?

Every living organism contains the wonderfully complex DNA molecule, yet has so many different arrangements that each species has unique sequences of DNA base pairs. DNA is a fantastic example of unity amidst diversity. Starr and Taggart believe "*This molecular constancy and variation among species is the foundation for the unity and diversity of life.*"[3]

Chemically, this is correct, but Scripturally, it is not. DNA is a "second cause." DNA is super-special and super-important, but it does not work by itself. DNA requires the correct environment containing the proper chemicals to unwind it and start the protein synthesis process. DNA does nothing without help from other chemicals, none of which will ever make or do anything without a First cause organizing them to produce and reproduce life.

> DNA does nothing without help from other chemicals, none of which will ever make or do anything without a First cause organizing them to produce and reproduce life.

Genetics

More terms to memorize!

a. **Genetics:** Study of heredity.

b. **Heredity:** The total genetic endowment obtained from the parents.

c. **Gene:** A segment of DNA containing the information for a heritable trait, passed on from parent to offspring.

d. **Allele:** Multiple forms of the same gene. Examples include alleles for hair color, eye color, and skin color.

e. **Chromosome:** A DNA-containing body. Found in the nucleus of human cells.

f. **Locus:** The exact location of a gene on a chromosome

g. **Homologous pair of chromosomes:** Two chromosomes having the same alleles in the same order. Human cells have 23 homologous pairs of chromosomes.

h. **Diploid:** The condition of having two pairs of chromosomes.

i. **Haploid:** The condition of having only one member of each homologous pair of chromosomes. A characteristic of egg and sperm cells (gametes).

j. **Genotype:** A single gene pair.

k. **Phenotype:** The physical trait expressed by a gene pair, or genotype.

l. **Homozygous:** Having the same alleles at the same locus on homologous chromosomes; a "purebred."

m. **Heterozygous:** Having different alleles at the same locus on homologous chromosomes; a "hybrid."

Gregor Mendel (1822-1884)

Most scientists consider Gregor Mendel, a contemporary of Darwin, the "father of genetics." Although we know little of his personal beliefs, we do know he was a Franciscan monk, and we know he rejected Darwin's ideas about evolution. His pea experiments, carefully controlled and mathematically analyzed, provided the basis for our understanding of heredity[4]. Mendel worked with thousands, as opposed to tens or even hundreds of pea plants, because he understood how important mathematical probability was in experimentation.

What is *mathematical probability*? We'll talk about it more in Chapter 4, but for now, here's a simple example. If I flip a coin, the mathematical probability of it landing on heads is 1 in 2, or 50%. If I test this with an experiment and flip a coin 4 times, I may or may not get the coin to land on heads 2 of 4 times. However, the more times I flip the coin, the greater the chances are that I will come up with results closer to the mathematical probability of 50%.

Mendel understood mathematical probability, which is why he studied so many pea plants. For example, in a test with pea seed color, he had 8,023 seeds in his experiment[5]! In any science experiment, the more times the experiment is repeated with similar outcomes, the more believable it is. Mendel hypothesized that plants might inherit two "units" (what we now call genes) of information about a trait, one from each parent. To test his hypothesis, he studied pea plants through two generations.

We can better understand Mendel's genetic tests by drawing a *Punnett Square* (Fig. 3.3). The Punnett Square helps us visualize the transfer of genetic material from parent to offspring. To study pea color, Mendel obtained plants that were homozygous (purebred) for yellow and green peas. He knew cross-fertilizing purebred green with purebred yellow plants would result in yellow pea plants, because the allele for yellow pea color dominates over the allele for green pea color. This is called *simple dominance*. Keep in mind not all alleles display simple dominance.

When using a Punnett Square, represent a dominant allele with a capital letter, and a recessive allele with a lowercase letter. For our discussion with peas, we will represent the dominant, yellow allele with a capital Y, and the recessive, green allele with a lowercase y. In a purebred yellow plant, all the cells in the plant have a pair of Y alleles. We say their genotype is YY, pronounced "big Y, big Y." When the plant forms gametes, or sex cells, which only contain one of each allele, the only possibility is for the gametes to have Y alleles. Likewise, a purebred green plant's genotype is yy or "little y, little y," and its gametes only contain recessive y alleles.

Using the Punnett square in Figure 3.3, it is easy to determine the genotype of offspring produced by crossing purebred green and yellow plants:

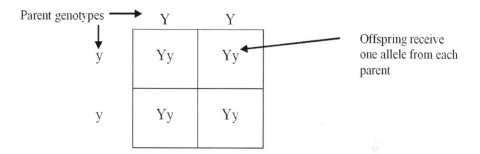

Figure 3.3. Punnett square for crossing of purebred green (yy) with purebred yellow (YY) peas.

The Punnett square gives the mathematical probability of the outcome of a crossing. In the cross shown in Fig. 3.3, there is a 100% probability the offspring will possess a Yy genotype, and therefore all peas would express the yellow phenotype. However, since Mendel couldn't see the genes, and since all the peas were yellow, he had no idea if they were YY or Yy genotypes. Therefore, he crossed the heterozygous Yy plants with each other, and studied this new generation, described in Figure 3.4.

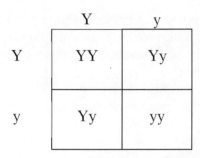

Figure 3.4. Punnett square for crossing of hybrid (Yy) pea plants.

3 out of 4, or 75% of the plants will have the yellow phenotype. A 50% probability exists for having offspring with a Yy genotype, and 25% probability for either YY or yy genotypes. In Mendel's actual experiment with pea color, he found 6,022 out of 8,023, or 75.06% of the plants were yellow, incredibly close to the predicted percentage.

The results of Mendel's pea experiments serve as the basis for the Theory of Segregation, stated as follows:

> *Diploid cells have pairs of genes, on pairs of homologous chromosomes. The two genes of each pair are segregated during the formation of gametes [reproductive cells], so they end up in different gametes.*[6]

Knowledge of mathematical probability is critical to understanding how traits pass from parents to offspring in sexually reproducing animals, and when experimental observations match up with theory, we gain confidence that what we are observing is really "the way it is."

The scientists of Mendel's day did not understand his experiments, and only later did others appreciate the significance of his work. Now, Mendel's experiments and theories serve as the foundation to any study of genetics.

Asexual reproduction

All the preceding genetics discussions referred to sexually reproducing organisms like birds, bees and bass. We have not discussed asexually reproducing organisms, like bacteria. These organisms only contain a haploid number of chromosomes, and when they divide, they make a copy of their DNA for the new cell. The two resulting cells are clones of each other. However, there are processes, such as *conjugation* in bacteria, where two bacteria exchange sections of DNA. Processes like conjugation ensure "unity amidst diversity" in asexually reproducing organisms. It is so important to look for unity amidst diversity in Creation, because it is an attribute that reflects the triune nature of God. He is the First cause, and DNA and reproduction are the second causes He employs to ensure His creation will always be a reflection of Him, leaving mankind without excuse for knowing about Him (Romans 1:20).

Chapter 3 Summary:

Your body is made from billions of cells, amazingly complex, tiny, self-replicating systems. Inside every single one of your cells is a nucleus, and inside that are 23 homologous pairs of chromosomes. The total number of chromosome pairs in a cell varies between organisms. Each chromosome is a continuous strand of DNA containing distinct sections called genes. DNA has the shape of a spiral staircase, and the steps are pairs of nucleotide bases linked in the middle by a hydrogen bond.

A gene's DNA contains a coded message for the manufacture of a specific protein. Proteins are just chains of amino acids, and every 3 bases on a DNA strand is a code word, or codon for one of the 20 known amino acids. Cells manufacture protein molecules via transcription and translation.

> Inside each cell, there are about 1 billion code words. This equals about 1000 books of 500 pages each.

The sum of all of your genes, or your genome, contains an astounding 3.2 billion base pairs of DNA., which means that inside each cell, there are about 1 billion code words. This equals about 1000 books of 500 pages each, using the smallest print you can read, and your cells can duplicate this information in about 8 or 9 hours!

By applying the truths of the senses, truths of understanding, and truths of belief to the question "How did DNA get here?", the only logical answer is that *the God of the Bible created DNA*. This truth becomes clearer when we study further the parable of the sower, quoted at the beginning of this chapter. Remember that parables are analogies or abstract truths applied to a variety of situations.

Dr. A.E. Wilder-Smith (1915-1995), a Bible believing Christian with three doctorate degrees, applied the parable of the sower to explain that God, not evolution, created the genetic code[7]. Wilder-Smith said that when we think, we think in a concept, and we put that concept into a code, which is our language. For example, if someone gives me a gift, I think of the concept of thanksgiving, and I put that concept into a code (the phrase "thank you"), and that code is my language. Smith said there is no natural law for codes, that they are entirely arbitrary.

For example, "gracias" (Spanish) also means thank you, and so does "spaciba" (Russian). There is no natural law that says when the concept of thanksgiving comes to my mind, the organization of that thought has to be expressed as the English phrase "thank you." The point here is that we can express the concept of thanksgiving by using a language, which is an organized code created by people. We use this code to organize and express our thoughts.

In the parable of the sower, Jesus told his disciples the seed represented the Word of God. Think about this now. From a material standpoint, what are seeds? Seeds are just "packets of genes," and genes are code words, and just like languages are expressions of human thoughts, the 64 code words contained in genes are expressions of God's thoughts! When we look at God's word, we see that when planted in fertile soil (a good and noble heart), the seed bears much fruit, but when planted in poor soil (a hard heart), it withers away. The same thing happens to a physical seed, and is an incredible example of God revealing His attributes to us through a study of His works. God is the First cause who created the "letters" of the DNA "alphabet," and used them to create assembly instructions for every living thing. The unity amidst diversity we see in Creation is an attribute of God, and the molecular constancy (every living thing has DNA) and variation (everything except clones have different arrangements of DNA) are because of Him, not in spite of Him.

4 MUTATIONS AND MATHEMATICS

Mutations and Probability

In Chapter 4 we will continue our study of biology fundamentals by covering genetic mutations and mathematical probability, and we will apply this knowledge to uncover an astonishing suppression of truth found in secular biology textbooks. We will end the lesson with a recap of the first 4 chapters. In the remaining chapters, we will apply our knowledge to discover more places where truth has been exchanged and suppressed in biology teaching.

We haven't formally defined evolution yet, so let's do that now. *Evolution* simply means "genetic change in a line of descent"[1]. The definition can be misleading unless you stop and think about what "change" is. Change can be good, bad, or neutral. For example, if I go fishing and catch 3 bass and then change lures, the change could be good (I'll catch more than 3 bass), bad (less than 3 bass), or neutral (3 bass again).

The following chapter seeks to answer two questions dealing directly with the formal definition of evolution. One is "What causes genetic change?" The answer is *mutations*, our first topic of study. The second question is "What causes mutations?" We have two answers to this question, *known causes* and *unknown causes*. Scientists use mathematical probability to study the unknown causes for mutations, which will be the second, and most important topic of study in the chapter.

Mutations

Starr and Taggart define a mutation as a "permanent alteration in a gene's base sequence"[2]. Mutations are the original source of genetic variation in populations. Changes brought about by mutations may lead to smaller or larger differences in traits within a population. The most common mutations are base pair substitutions, insertions, and deletions[3]. Substitution mutations cause 1 amino acid to change, so it changes 1 code word in a gene's base sequence. It would be like writing the sentence "*Thou madest him to have dominion*" as "*Thox* madest him to have dominion." One misspelling, yet the meaning is still discernible.

> Mutations are the original source of genetic variation in populations.

Insertions and deletions change part or all the sequence, depending on their location. For example, inserting an "A" at the beginning of the above example, and keeping the number of letters in each word the same, would result in "ATho umades thi mt ohav edominio." Insertion and deletion mutations create many more illegible codes than do substitution mutations.

Causes of mutation

Unknown causes: (a.k.a. "spontaneous" mutations). Each gene has a characteristic mutation rate, which is the probability that it will mutate spontaneously. For example, in a single reproductive season, about 1 in 100,000 to 1,000,000 human gametes has a new mutation at a given locus[4]. In another example, the probability of an E.Coli bacteria's gene mutating in such a way that it develops a resistance to streptomycin (antibiotic) is 1 in 2.5 billion cells[5]. That sounds like a very low probability, unless one considers there are roughly 4 billion bacteria in a single capsule of acidophilus, a helpful bacteria some people consume to aid digestion. Put in perspective, it does not seem that "impossible" for bacteria to develop antibiotic resistance, even though the probability sounds so low. Keep in mind though that all known mutations for antibiotic resistance are the result of a loss, not a gain of genetic information.

Known causes: Mutations of known cause can result from mutation-causing agents or mutagens[6]. Radiation, as well as natural and manufactured chemicals can alter DNA in various ways. Many of these are cancer-causing agents, or carcinogens. Cancer is just a mass of mutated cells that divide abnormally, and many known causes, or carcinogens, exist, including everything from artificial sweeteners to second-hand smoke.

Transposons are another mutation of known cause. Transposons are segments of DNA that can move to different locations in a genome. They may inactivate genes into which they insert themselves and cause changes in phenotype. Barbara McClintock discovered them in the 1950's while working with corn (Fig. 4.1). The discovery was in response to a question she had about why some kernels of Indian corn were dark and other kernels were yellow. She discovered that transposons were causing the gene for color production to turn off in some kernels, making them yellow.[7]

Figure 4.1 Transposons are segments of DNA that can turn some genes on and off. One example is the color-producing genes in corn.

Other facts about mutations

- In sexually reproducing organisms, the only mutations affecting off-spring are ones occurring in the gametes of the parents. If mom has a stomach cell that gets a mutation, the mutation will not pass to her child. However, if she has an egg cell with a mutation, it will pass to her child.

- A protein created by a mutated gene may have a harmful, neutral, or beneficial effect on the organism. Keep this in mind, and remember the definition for evolution is "genetic change."

- A 1990 Genetics textbook states "Most mutations will be detrimental ... a few beneficial"[8], and in a 1990 Biology textbook "Mutations are generally harmful, on rare occasions beneficial"[9]. Contrast these statements with the 2004 edition of Starr and Taggart: "many mutations are harmful, some are harmless or beneficial"[10]. Most mutations are harmful, but as the 2004 biology textbook reveals, the truth about mutations is suppressed by using more vague words. "Many harmful, some harmless" does not sound nearly as extreme as "most harmful, few beneficial." Why the change? I think it is because there is a suppression of truth going on regarding beneficial mutations, which I am about to discuss. Stick with me for the rest of the chapter, and you can decide for yourself why descriptions of mutations have changed.

Why is it important to know about gene mutations?

Since evolution means "changes in genes," and the only way a gene can change is by mutations, then one must deduce that evolution occurs by gene mutation. Page 10 of Starr and Taggart says that mutations are "the original source of variations in heritable traits"[11]. If we use this statement as our major premise, then the conclusion is that all the species could be the result of mutations. This is a logical argument, but as we have learned, just because an argument is logical doesn't mean it is true.

Put another way, the major premise of secular biology textbooks is that evolution in the form of "new and improved" phenotypes occurs, and I believe this is a false major premise.

Mutation Summary:

Gene mutations can be harmful, beneficial, or have no effect. Mutations occur by either known or unknown causes. Transposons generate mutations that do not harm the organism, and mutagens predominantly generate harmful mutations. All so-called "beneficial" mutations generate only from unknown causes. Starr & Taggart's 2004 biology text gave no examples of beneficial mutations arising from known causes. If there were some examples of the causes of beneficial mutations, they would be discussed in greater detail.

In order for an animal to change into something else, it must have a bunch of beneficial mutations. Beneficial mutation would allow it to gain an advantage over its competition or to fill a position or "niche" in the ecosystem. Harmful mutations lead to death, and neutral mutations just create variety within a species, which leaves beneficial mutations as the only mutations that could possibly create new species. However, we don't know how, or for the most part if, these beneficial mutations arise. We just call them "spontaneous" mutations, but spontaneity does not cause anything. Transposons, radiation, and carcinogenic chemicals cause mutations, but spontaneity cannot cause anything. It is not a material thing. We know that most genes mutate spontaneously at measurable rates and are understood by using mathematical probability, but mathematical probability does not cause anything either. Nevertheless, the whole idea of evolution of new species rests on the probability that new, beneficial genes will form spontaneously. To understand this more, let's dive deeper into the fascinating world of mathematical probability.

Probability

We discussed probability, or *chance* in Chapter 3 while covering genetics and Punnett squares, and while covering spontaneous mutation rates in the current chapter. Now, we will present a more formal definition. Mathematical probability is just a fraction, so if you understand fractions, then you can understand probability. A common formula used to represent probability is this:

$$P = (\text{number of desired outcomes})/(\text{number of possible outcomes})$$

To better understand probability, let's review our coin-flip example from Chapter 3. If I were flipping a coin and I wanted it to land on heads, then heads would be the *desired* outcome. Only two *possible* outcomes, either heads or tails, result from a coin flip. Therefore, the probability of a flipped coin landing on heads is1 over 2 or ½. You can express the probability as a fraction (1/2), a decimal (0.5), a percent (50%), or a ratio (1:2 or 1 in 2).

Think about the meaning of the word *probability*. There is a reason the idea

> Mathematical probability helps one predict the outcome of an event, but it never predicts with absolute certainty what will happen.

it describes is not called "definitely" or "mathematical exactness," or "absoluteness." For example, if I flip a coin 100 times, I will *probably* get close to 50% heads, but I *probably* won't get heads exactly 50% of the time.

Do scientists use mathematical probability? Yes! We have already looked at two situations where scientists use it: Punnett squares and rates for spontaneous mutations. Mathematical probability helps one predict the outcome of an event, but it never predicts with absolute certainty what will happen. Nevertheless, it is a useful tool that helps scientists accept or reject hypotheses.

Does God use mathematical probability?

I don't know if He does or not, but He definitely allows us to use it as a way to reveal Himself and His will in a way we can understand. Proverbs 16:33 says "the lot is cast into the lap, but its every decision is from the Lord." "Lots" is a game of probability, or chance, kind-of like drawing straws. There are several examples in the Bible of casting lots to determine God's will (Jonah 1:7, Acts 1:24-26).

For spontaneous mutations, which are mutations of unknown cause, maybe there is a second cause driving them and maybe there isn't. Whatever is happening, we can be sure the God of the Bible, who abhors inaccurate quantities (Proverbs 20:10) and counts even the hairs on your head (Matthew 10:30), reveals Himself to us with numbers. Therefore, we should not be surprised that He might use probability to reveal Himself to us. Mathematical probability doesn't cause anything, and in the words of theologian R.C. Sproul "as a causal force, chance remains, ever and always, a fiction"[12]. Probability doesn't cause anything, but God can, and if he wants 1 in every 2.5 billion E.Coli cells to mutate spontaneously, then He can!

Are we going somewhere with this?

YES! My hope is to try to show you that probability is a useful tool for the study of God's works and the revelation of His will. We have seen its usefulness in determining genotypes and phenotypes of offspring using Punnett squares and we have seen how we can use it to predict rates of spontaneous mutations.

There is one topic of unprecedented importance that we haven't looked at yet, so if you are still a little shaky on what probability is, then you might want to stop and review. The topic we are about to discuss is found nowhere in the biology textbooks, because to discuss it would turn evolution into a meaningless, pointless theory. We have talked about some exchanges of truth found in biology textbooks, but this topic is not even mentioned, and therefore it is a suppression of Truth (Romans 1:18).

If you want to reject what I am about to say, that's fine, but realize the concept is based on mathematical probability. Understand that if you deny this, then you must deny that genes spontaneously mutate at known rates, that Punnett squares don't predict anything, and that all mathematical probability is useless in science. What I am talking about is *the probability to create new genetic information*.

The probability to create new genetic information

We have learned that evolution means *changes in genes*, and the source of changes is mutations, and mutations can be good, bad, or have no effect. Good mutations, as far as scientists know, only arise spontaneously, or from unknown causes. If we really did evolve then at one time we were simple bacteria with only 4,000 or so genes, and now we are humans with 30,000 genes in our genome. Therefore, we had to get at least 26,000 new genes from somewhere, all of which benefited the previous species, yet none of which do we know how it got there. Hmmm....

To understand the probability of new genes forming, think about a simple gene only 10 base pairs long, and think about how you could determine the probability of spontaneously forming it in a certain order. Remember that probability is the ratio of desired outcomes to possible outcomes, and in this example there is only one desired outcome that we want. But we still need to find the number of possible outcomes.

We can find the number of possible outcomes using a mathematical procedure called a *permutation*. Here's how permutation works. If I went to a fast-food restaurant and saw the menu listed 5 hamburger choices, 4 drink choices, and two sizes of fries, I could do a permutation to find there are $5 \times 4 \times 2 = 40$ burger/drink/fry combinations to choose from.

Now, get out a paper and pencil and draw four connected squares, big enough to put a number into each. Think of each square as a DNA base pair. There are 4 different bases (adenine, cytosine, guanine, and thymine), so there are 4 choices for the 1st square, 4 for the 2nd, etc. Following the restaurant example, there would be 4x4x4x4=256 possible base pair combinations in this little DNA strand.

Next, draw 10 squares, representing our 10 base-pair DNA example, and put a 4 inside each square. For the whole 10 base-pair strand, there are 4^{10}, or 1,048,576 possible combinations. That means there is about a 1 in 1 million chance that this little 10 base-pair strand of DNA would form spontaneously in a certain order!

Put another way, if I gave 1 million people a large bag of base pairs from a DNA plastic model, blindfolded them and asked them to assemble 10 of the pairs in a certain order, I could predict that only 1 person would assemble it in the order I wanted! However, as we have learned with studying bacteria cells, numbers in the millions and billions are not really that big, so let's apply what we just learned to a real gene. In Chapter 3, we learned the average gene in human chromosome 22 was 19,000 base pairs long. The probability of spontaneously forming our 10 base-pair gene is 4^{10}, so hopefully you can see the probability of spontaneously forming a 19,000 base-pair gene is 1 in 4^{19000}, which is about 1 in 1- with 11,439 zeros! Now we are talking massive numbers, inconceivably large.

Right now you may be saying "well, there's still a *chance* it could form," and yes there is. If you want to believe that something with that low of a probability could form then you can if you want to, just like you can lie if you want to and you cannot believe in God if you want to.

Think about one more thing though. In our probability examples for creating a plastic DNA model, these are the probabilities *with the help of someone's hands putting the base pairs together*. In a cell, what are the "hands" that help replicate DNA, make RNA, and synthesize proteins? The hands are the enzymes. The probability of forming DNA "without hands" would be about the same as putting plastic model base pairs in a Ziploc bag, shaking them every once in a while, and eventually expecting 10 base pairs to stick together. The probability of that happening is so low it is not worth discussing.

Before we finish, let's look at one final example. Scientists have determined[13] the minimum number of protein molecules needed to theoretically form a living organism would be about 239. The probability of these forming is an unbelievably small 1 in $10^{119,879}$. And the time it would take for these 239 proteins to form is $10^{119,841}$ years[14]! Current evolutionary theories estimate the earth at 10^9 years old, while Biblical ages are between 10^3 and 10^4 years. Either way, whether you believe the Biblical age of the earth or theoretical age, there just isn't enough time for spontaneous mutations to be the cause of life on earth.

Now what?

We cannot call this last section an exchange of truth, because biology books don't even discuss it. We can call it a suppression of truth though, which goes along with Romans 1:18-25. This passage begins "For the wrath of God is being revealed against all the unrighteousness and ungodliness of men, who *suppress the truth in unrighteousness….*" What this passage is saying is that people who are "not right" will suppress the truth.

People write biology textbooks, so people are being unrighteous by not including the probabilities of creating new genetic information in biology textbooks. We just proved mathematically that there is a 100% probability life did not evolve spontaneously, and a 0% chance that it did! I can comprehend 1 in 1 million, and even 1 in 1 billion, but 1 in $10^{119,879}$ is as close to

> We just proved mathematically that there is a 100% probability that life did not evolve spontaneously, and a 0% chance that it did!

"not a chance" as we can get. If you want to cling to that one chance, then you can do that, but keep in mind that all of these probabilities still require "hands" to make them work. It takes hands to flip coins, it takes hands to assemble plastic DNA models, and it takes hands in the form of enzymes to form real DNA.

Chance is nothing. It cannot, nor will it ever "cause" something. When we look to the First cause, the real "hands" that put it all together, we see that *"All things were made through Him, and without Him nothing was made that was made"*(John 1:3).

Chapter 4 Summary

Charles Darwin is the founder of the current theory of evolution. Biology textbooks, such as p. 3 of Starr and Taggart, clearly state *"Theories of evolution, especially the theory of evolution by natural selection as formulated by Charles Darwin, help explain life's diversity."*[15]

Darwin was unaware of many of the things we almost take for granted today. He knew there were problems with his theory, and he considered the lack of transitional forms in the fossil record to be "the most obvious and gravest objection which can be urged against my theory"[16]. But there was an even "graver" problem that he did not consider, and it is one that he could not consider. Darwin did not know about chromosomes, genes, DNA, or nucleotide bases. He did not know that mutations are the source of changes in genes and that most mutations are harmful.

Darwin did not like math very much either and considered his impatience with learning even the basics of algebra "*foolish,*" and something that he "*deeply regretted…. for men thus endowed seem to have an extra sense*"[17]. We have used that "extra sense" to reveal the probability, or actually the *im*probability of life evolving spontaneously "by chance."

Darwin had no idea how complex cells are. He had no concept of transcription and translation, or of the coded message contained on the strands of DNA

> Darwin had no idea how complex cells are.

that contain instructions for making you and me. He could not, therefore, understand the total impossibility of making a beneficial gene, much less any life at all "without hands." Yet we still discuss his theory as if it were the unifying concept in biology. Why do we do this? We can look to God's word, and find the answer in verses like Romans 1:18-25 and Colossians 2:8. Human beings are easily deceived by philosophies based on the traditions of men, rather than on Christ.

Francis Bacon, the founder of the scientific method, despised learning styles based solely on the traditions of man. He quit college after two and a half years because he found "all studies were confined to Aristotle, who was considered infallible in philosophy, a dictator to command, not a consul to advise," and "the studies of men…. are pinned down to the writings of certain authors; from which, if any man happens to differ, he is presently reprehended as a disturber and innovator."[18]

History often repeats itself, and it should be no surprise to see the same problems Bacon described in the educational system of his day occurring again, over 430 years later. This time, many confine their studies to Darwin, considering those who differ from Darwin's philosophy as "disturbers and innovators." Bacon said this type of learning was analogous to a man in a large room using a single "watch candle" and only being able to see one portion of the room[19]. Trusting Darwin's philosophy and studying him alone is like trying to study a large, dark room with one little candle that's too dim to make much sense of anything.

If you don't follow God, you will replace God with something. God has designed you to know Him, because *what may be known of God is manifest in you* (Romans 1:19). He has designed you to believe, and if you don't believe in Him, then you will make something else God, either yourself, another person, or some other created thing.

God has a plan for you as well, whether you believe him or not. If you believe Him, he will give you eternal life (John 3:16). If you don't believe Him, he will give you up to the lusts of your hearts (Romans 1:24), which means all you will think about is yourself, what you can get for yourself, and a constant search to find something or someone to worship and serve. If that describes you, all you have to do is "repent and be baptized in the name of Jesus Christ for the remission of sins; and you shall receive the gift of the Holy Spirit. For the promise is to you and to your children, and to all who are afar off, as many as the Lord our God will call" (Acts 2:38-39).

In this chapter we used our skills of deductive reasoning by applying mathematical probability to the study of mutations. We also used our thinking skills by asking questions about the probability of forming life. In his book, *The Wedge of Truth*, Phillip Johnson summarizes a study of mutations and probability by saying *"information-creating evolution is not empirical science at all because it has never been observed either in the wild or in the laboratory"*[20].

> Information-creating evolution is not empirical science at all because it has never been observed either in the wild or in the laboratory
> *Phillip Johnson*

Are you holding tight to the "watch candle" of Darwinism, which dimly lights one corner of the room while obscuring the rest in darkness? Are you possibly responsible for promoting this "disease of learning," as Bacon called it, by teaching biology or writing biology textbooks with this one man as the foundation? Darwin, like Aristotle, had some good ideas and some bad ideas, and if we are really serious about advancing learning, we will throw out the bad ideas and keep the good ones. In the upcoming chapters, we will take a detailed look at Darwin's theory, as well as Biblical creationism, and intelligent design. We will ask serious questions about each of these, all the while keeping our major premise in mind that science and religion do mix, and when we separate them, we become "futile in our thinking" (Romans 1:21).

PART III

Evolution, Creation, and Intelligent Design

5 GENOVERSITY
Microevolution

Secular biology textbooks typically break any study of evolution into two chapters, microevolution, and macroevolution, including a discussion on speciation. *Microevolution* refers to genetic diversity within a species, while *speciation* refers to the possible results of microevolution and *macroevolution* refers to genetic diversity between species. We will discuss microevolution presently and macroevolution in Chapter 6.

In Chapter 4 you learned that mutations could be good, bad, or have no effect on the organism. To begin our study of microevolution, we must first polish our definition of genetic change. Most genetics textbooks refer to mutations as either gain in function mutations or loss of function mutations, and these can have varying degrees of good or bad effects on the organism. In loss of function mutations, the product (the protein produced by the gene) has a reduced function or no function at all. These usually cause recessive phenotypes. In gain of function mutations, the product is something "positively abnormal"[1], such as an increase in production of a growth hormone, which results in a taller height. Gain of function mutations usually produce dominant phenotypes. There are also mutations that cannot be classified as gain or loss of function mutations.

Basic Mendelian genetics, where we talk about "Big *A*" and "little *a*" alleles, is helpful in understanding the fundamentals of genetics. However, to understand genetics in greater detail, we must quickly realize that for most genes, there are a multitude of alleles (remember that alleles are multiple forms of same gene). For example, when describing the genotype of a cystic fibrosis (CF) carrier as *Aa*, what we really mean by the recessive *a* gene is any gene that has mutated so it does not produce a functioning chloride channel. Currently over 750 such alleles have been documented. Similarly, *A* means any functioning sequence, and as long as the person has at least one of the *A* alleles, they will not get cystic fibrosis because they will have the ability to produce a functioning chloride channel[2].

CF, a disease studied in great detail, has provided for us a window into the incredible genetic diversity found in populations. When we think of genetic mutations as either resulting in a gain or loss of function, it becomes apparent there are various degrees of "goodness" and "badness" that result when the genes are expressed. This diversity within the "gene pool" of a population is the result of the effects of what scientists commonly refer to as *microevolution*.

Microevolution is defined as the *change in allele frequencies in a population*[3]. For example, in a population study of fish called arctic grayling, if 2% had 13 spots on their side, and then 1 year later the same population was studied and 3% had 13 spots, we might conclude that microevolution occurred, because we observed a change in the frequency of the gene for creating thirteen spots. But what caused the change? The four main causes discussed in biology books are *mutations, natural selection, genetic drift, gene flow*, or some combination of these.

Mutations

Mutations drive genetic change. Mutations alone create new alleles[4], while natural selection, genetic drift and gene flow just change the frequency of occurrence of alleles in a population. Keep in mind that all mutations of known cause are harmful to the organism or cause no effect, and there are no known mutations that create new, beneficial genetic information.

Natural selection

Natural selection is *the outcome of differences in survival and reproduction among individuals that differ in details of heritable traits*[5]. Charles Darwin first proposed the idea in his 1859 book *On the Origin of Species*. Darwin possessed many attributes that would hamper his understanding of science and the scientific method (he wasn't very good at math[6], he didn't like to dissect or draw[7], and he despised deductive reasoning[8]). However, he had one important attribute no scientist can be without. He was incredibly observant. His observational skills allowed him to see minute differences in phenotypes within a species. He made countless observations of "variation under domestication," which is what we might refer to as *breeding* or *artificial selection*.

There is no doubt, when we look at all the different breeds of cats, dogs, etc., that we can select for certain desirable traits within a species. Darwin applied what he learned from breeding, as well as what he learned from Thomas Malthus (1766-1834), to develop his theory of natural selection.

In most studies of Darwin's theory of natural selection, little is said of Thomas Malthus. However, he greatly influenced Darwin's thinking. Malthus was an English political economist and historian who, in 1798 published a book called *An Essay on the Principles of Population*. He proposed that poverty, and thereby also vice and misery, are unavoidable because human population growth will always exceed food production. He used some fancy graphs showing how populations might increase with a *geometric* progression pattern (1,2,4,8,16,....), while food production only increases in an *arithmetic* progression pattern (1,2,3,4,5, ...). The checks, or controls on population growth were wars, famine and diseases.

While Malthus never managed to put forward any scientific proof for his theory, it nevertheless promoted repressive legislation which worsened conditions for the poor in England. England's leaders reasoned that better conditions for the poor would only encourage further propagation, putting those who were capable of work at a disadvantage. In 1834, England created workhouses for the poor, and the sexes were strictly separated to curb the otherwise inevitable over

> Darwin took the false doctrine of Malthus and made it a cornerstone of his own theory.
> *Bernhard Schreiber*

breeding. Obviously, this type of thinking has an inherent devaluation of human life through fear that the ever-increasing population of lower classes will devour the more civilized or "better" people[9].

During his research Darwin came across Malthus' essay and realized he could expand his own theory to include all life in the inevitable struggle for existence if food production were to lag behind population growth. Darwin took the false doctrine of Malthus and made it a cornerstone of his own theory[10]. In his own words, Darwin said

> "The Struggle for existence amongst all the organic beings throughout the world, which inevitably follows from their high geometrical powers of increase, will be treated of. This is the doctrine of Malthus, applied to the whole animal and vegetable kingdom. As many more individuals of each species are born than can possibly survive; and as, consequently, there is a frequently recurring struggle for existence, it follows that any being, if it vary however slightly in any manner profitable to itself, under the complex and sometimes varying conditions of life, will have a better chance of surviving, and thus being naturally selected. From the strong principle of inheritance, any selected variety will tend to propagate its new and modified form"[11].

Darwin used an analogy to show that in artificial selection, man is the selector of desirable traits, and in natural selection, "nature" selects the desirable traits. Darwin thought nature selected the best or fittest individuals to survive and reproduce, the result being adaptation to the environment.

Darwin based his definition of natural selection on the idea that animals and plants would be in competition. There is no doubt that competition for food, territory, mates and other resources occurs within and between populations. However, the results of this competition do not necessarily result in the "fittest" individuals surviving.

For example, male Northern Fur Seals arrive on the shores of Alaska's Pribilof Islands in early summer, and battle each other for territories along the shore (Fig. 5.1). The males who win a territory will care for a group of females and their young.

The reasons for a male's victory may not necessarily result from good genes. Perhaps, the losing seal had a broken flipper from a previous accident, allowing an easy victory for the healthy seal. The loser may have a better genetic makeup, but lost nevertheless and was unable to pass his better set of genes along.

Figure 5.1 Young fur seals play-fighting, Pribilof Islands, Alaska. The animal with the best genes doesn't always win.

Another example is rainbow trout feeding on king salmon eggs. Rainbow trout lurk downstream of schools of spawning salmon, waiting patiently for eggs that for one reason or another wash from the salmon's nest. The genetic makeup of the baby salmon inside the egg has no bearing on whether a trout eats them. It would be ridiculous to think that rainbow trout were "naturally selecting" the weakest genetic stock of salmon. Countless other examples exist of competitions where the "best" were not the victors. When a predator seeks out its prey, it is not thinking "I'm going to eat you because you are genetically inferior," it is thinking "If I can catch you, you're dinner."

To conclude, while natural selection does occur and can result in offspring being better adapted to their environment, it doesn't necessarily mean they received the best genetic material. Moreover, since mutations are usually harmful, we can reason that "better adapted," which can only occur because of mutations, doesn't mean "better genes." An excellent example occurs with antibiotic resistant bacteria on the video "Icons of Evolution"[12].

Gene flow

Gene flow is the physical movement of genetic material into and out of a population[13]. For example, transfer of genetic material occurs during pollination of a population of wildflowers by a group of insects from another location. (Fig 5.2). The result could be either a more-fit or less-fit wildflower population, or there could be no change.

Figure 5.2. Wild geraniums, Katmai National Park, Alaska. If insects from another location pollinate these flowers, gene flow occurs.

Genetic drift

Genetic drift is the change in allele frequencies in a population over many generations, brought about by chance alone[14]. In Chapter 4 we discussed how chance can't do anything, so chance can't cause the frequency of alleles to change in a population. What really happens is that, for whatever reason, a previously large population becomes small, and it then has only the genes from this small population to work with.

There are two main ways a population becomes small. One is when there is a severe reduction in population size brought about by disease, habitat loss, hunting, or some other natural disaster. The population only has a few animals from which to rebuild. This is called a bottleneck. A good example is the animals on the ark. From what Scripture reveals, there were no fishes on the ark, but all other animal groups were present. The second way a population becomes small is when a few organisms leave a population and found a new one. Birds or plants inhabiting a newly formed island are a common example. A school of salmon reproducing in a newly-formed stream is another example (Fig. 5.3).

Figure 5.3. Sockeye salmon, Russian River, Alaska. Genetic drift can occur when salmon populate a newly formed stream.

More thoughts about microevolution

Is *microevolution* a good word for describing genetic diversity within a species? No, it's not. While textbooks may say that evolution simply means genetic change, and that change can be good, bad, or have no effect, we all know that when somebody says this or that "evolved" they mean that it is "new and improved." In our society, we see the word "evolution" used in everything from naming cars to discussions of the "evolution" of education programs. The key issue is information, and if new genetic information

> A good word to replace microevolution is genoversity. This word is a combination of two words; "genome" and "diversity."

was created that benefits the organism, there is the possibility new species will form as a result. As we learned in Chapter 4 however, new genetic information benefiting an organism has never formed.

A good word to replace microevolution is *genoversity*. This word is a combination of two words; "genome" and "diversity." Scientists have proven that diversity exists within the genome of a species, and even someone with poor observation skills can tell there is diversity within populations of organisms. Just take a look in the mirror, and then go look at your parents, and you'll see both similarities and differences between you. While food quantity and quality and other environmental factors create some of the diversity, much of it results from genetics.

There are *limits* to the diversity of alleles for a particular gene. In loss of function mutations, the mutations can cause anything from a slight loss of function to a complete loss of function of the gene. Complete loss of function many times leads to disease and/or death of the organism. In gain of function mutations, a slight gain in function may be beneficial to an organism. For example, if the elastin gene has a gain in function mutation, this may allow the organism to have better-than average tissues that require elastin (skin, lungs, heart).

Gain of function mutations have also created genes with a novel function. This means that we know of genes that have mutated and created new information. However, a closer look reveals that these genes are pathogenic, and are most often noticed when they lead to cancer[15]. The book, *Human Molecular Genetics* states that gain of function mutations were "no doubt important in evolution"[16]. However, no examples exist in this text or any other of a gain of function mutation producing a gene with a *beneficial*, novel function. If evolution were true, examples of new, beneficial genes should be everywhere. But there are none.

The Peppered moth story was faked. On page 10 and 283 of Starr and Taggart, and on the pages of many other biology texts, the story of the peppered moth is used to prove Darwin's theory of natural selection. The story begins in the early 1800s, when almost all peppered moths were supposedly light colored, but during the industrial revolution, their colors shifted mostly to black. According to evolutionary theory, they were better-camouflaged against pollution-darkened tree trunks and were therefore better able to survive against bird predation. However, when the pollution diminished in the 1960's, the light-colored moths made a comeback.

The fact that moths began to turn light before the tree trunks did puzzled scientists. It turned out that the photographs of moths on tree trunks had been staged. Afterwords, several decades of research involving tens of thousands of moths revealed that only 6 were found resting on tree trunks[17]. This story is still in biology textbooks and is a suppression of truth, because articles in newspapers and scientific journals exposed the fallacy back in 1998.

Diversity within populations is an attribute of God. *"For as we have many members in one body, but all the members do not have the same function, so we, being many, are one body in Christ, and individually members of one another."* (Romans 12:4-5). Scientists believe the human body alone has over 200 different types of cells, which is amazing, especially considering all cells have identical chromosomes. On a different scale, a study of any group of people will immediately reveal to you there is great diversity within even the smallest population. In Romans 1:20, God said that His invisible attributes are clearly seen, being understood by the things that are made, so men do not have an excuse for not knowing about God. We have talked about the unity amidst diversity concept and we cam definitely see God's triune nature when we observe the diversity existing within the population of any species. God has obviously put instructions into genes to insure diversity will occur, although not always perfectly, which is a result of man's sin (Genesis 3:17).

If microevolution is true, why do we need the Endangered Species Act? There are many species listed as "endangered" or "threatened." The present state of any given endangered species closely parallels the "bottleneck" description of genetic drift mentioned earlier. However, on page 289 of Starr and Taggart, they describe bottlenecks, and the resultant inbreeding that occurs, as a "bad combination for endangered species."[18]

> If microevolution really worked, we would be eagerly anticipating the formation of "new and improved" species.

But how can bottlenecks act as both sources of microevolution, as well as "bad news" for populations? If microevolution really worked, we would be eagerly anticipating the formation of "new and improved" species resulting from the bottleneck in an endangered species' population. The fact that we don't is a powerful argument against all forms of evolution.

The more I read of Darwin, the more convinced I am that he was blinded by his desire to "show that the species had not been separately created"[19]. To develop his theory of natural selection, Darwin applied what he had learned from breeding and from Thomas Malthus' population essay. Darwin wanted to find a way to explain "the innumerable cases in which organisms of every kind are beautifully adapted to their habits of life, for instance, a woodpecker or tree frog to climb trees, or a seed for dispersal by hooks or plumes"[20]. Upon returning from his worldwide voyage on the HMS Beagle, he began extensive research and fact collection on domesticated plants and animals, concluding that "selection was the keystone to man's success in making useful breeds of animals and plants. How selection could be applied to organisms living in a state of nature remained for sometime a mystery."[21] Darwin solved his "mystery" in 1838 when he read Malthus' *Population* essay.

After reading Malthus' essay, it struck Darwin that under the circumstances of a struggle for existence "favourable variations would tend to be preserved, and unfavourable ones to be destroyed."[22] Now, let's stop here and question Darwin's theory. First, why did Darwin look to breeding programs to understand variation in populations instead of just looking to nature? Because he couldn't find any useful examples in nature.[23] In artificial selection, man is driving the selection by regulating the breeding. Man does not always select for the strongest, but sometimes he wants a certain color, shape or size. In the same way, in natural selection, in reality the "best" don't always survive, as in our examples with the fur seals and salmon.

Darwin admitted that his math skills were poor, and the major point of Malthus' *Population* essay revolved around a mathematical graph that he made of population growth compared to food production, which had no scientific data to support it. Darwin's bias against "separate creation" as well as his poor math skills probably contributed to him blindly embracing Malthus' make-believe graphs, and we are reaping the consequences of his lack of critical thinking. Now, I am not saying natural selection never happens, but it certainly doesn't cause *all* surviving offspring to be "better fit" than their parents, and it really doesn't explain much about how animals became adapted to their environment. There are many awesome videos worth watching that discuss evidence against natural selection as a driving force in nature, such as *Icons of Evolution*, and any of the videos by Dr. Jobe Martin (*Incredible Creatures that Defy Evolution*).

Chapter 5 Summary

Gene mutations are the driving force behind diversity within populations. Genes either gain or lose function because of mutations, and extreme cases result in disease and/or death. Mutations have even created genes with novel functions, but the genetic products created either killed the organism or caused cancer. We should expect diversity within a population, because this is an attribute of God.

We know there are limits to genetic change. Darwin's theory of natural selection, based on his observations of artificial selection and the false doctrine of Thomas Malthus, says that population growth will always outpace food production, and a struggle for existence will ensue. It is interesting that in Starr and Taggart's text, Thomas Malthus is discussed (p. 276), but no mention is made of his graphs being based on make-believe data [24]. This is yet another suppression of truth found in biology textbooks.

Darwin used his theory of natural selection as an attempt to explain how animals become adapted to their environment. He tried to use the theory to explain the origin of new species on the earth as well. We call this *macroevolution*, and we will discuss it in Chapter 6.

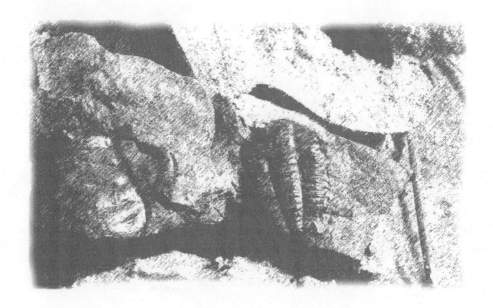

6 NO PROOF
Macroevolution

Page 3 of Starr and Taggart's secular biology textbook states that "Theories of evolution, especially the theory of natural selection as formulated by Charles Darwin, help explain life's diversity. The theories unite all fields of biological inquiry into a single, coherent whole"[1]. In Chapter 5, we learned that Darwin based his theory on artificial selection (breeding) as well as Thomas Malthus' population theory.

We know artificial selection is a real process, and I have no problem with it. I think it is one way God reveals to us the truth that he wants us to manage His creation, to "rule over" it, as He says in Genesis 1:28. What I do have a problem with is Malthus' population study, because it is not supported by any real scientific data. Recall that Malthus said "Population, when unchecked, increases in a geometrical ratio. Subsistence increases only in an arithmetical ratio."[2]. The graph in Fig. 6.1 is an example of his theory.

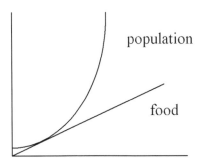

Figure 6.1 According to Malthus, population growth (curved line) always outpaces food production.

As "evidence" for this trend, Malthus looked to the United States, where, in the late 1700's, the population doubled about every 25 years[3]. I don't know if this data was true, but even if it was, the population increase was largely due to immigration, not reproduction within the existing population. Nevertheless, Malthus used this as his "rule" for population increase. One would think he would use concurrent data on food production in the United States to discuss subsistence, but

> A good scientific theory accurately describes a large class of observations, and makes definite predictions about the results of future observations.
> *Stephen Hawking*

instead he makes up a situation. The fallacy of Malthus' argument reveals itself when we look at food production today. In America, our farmers are so good at producing food the government actually pays some farmers NOT to grow crops!

It just amazes me that *Darwin did not base the theory of natural selection on any real evidence from nature, but rather on breeding programs and make-believe graphs!* Stephen Hawking, one of the most famous scientists alive today, describes a good scientific theory as one that accurately describes a large class of observations and makes definite predictions about the results of future observations[4]. Darwin's theory of natural selection does neither very well.

Mutations, genetic drift, gene flow, and the effects of the environment on these processes best describe genetic diversity within populations. The parable of the sower discussed in Chapter 3 is an excellent analogy of the effect of the environment on genetic information (Luke 8:4-15). Now, let's turn from genetic diversity *within* populations to genetic diversity *between* populations. Keep in mind that an evolutionary view of life proposes that all "species that have ever lived are related by descent"[5]. Since macroevolution stems from microevolution, which stems from fallacy and not real observations, we can expect more of the same. As you will see, the incredible (incredible meaning *not* credible) arguments found in macroevolution are so numerous they are difficult to keep up with. In this chapter I will list some of the major points of macroevolution theory, discuss flaws in each, and end with an important conclusion.

Speciation

Most secular biology texts describe evolution in three steps; microevolution, speciation, and macroevolution. Scientists define *speciation* as "changes in allele frequencies that are significant enough to mark the formation of descendent species from ancestral ones." Speciation describes a potential outcome of natural selection and other microevolutionary processes[6]. It is the predecessor to macroevolution.

In speciation, we have the supposed formation of new species. But what is a species? According to current biology textbooks, species are "groups of interbreeding natural populations that are reproductively isolated from other such groups,"[7]. This is the definition created by famous evolutionary scientist Ernst Mayr of Harvard University.

Evidences for speciation.

Chapter 18 of Starr and Taggart gives several examples of speciation. All the examples merely display slight morphological differences, with color patterns being one of the main differences. In addition, the textbook reveals that 4 of the 7 examples still interbreed, which, according to Mayr's definition, doesn't make them species. There is one example (Fig. 18.2) that does look like different species of plants, but it is actually the result of differing environmental conditions and not differing genetic makeup! Their "evidences" are simply examples of genetic drift and/or hybridization.

What about asexually reproducing organisms?

Page 294 of Starr and Taggart says that "the species concept does not apply to asexually reproducing organisms, such as bacteria"[8]. Why not? Well, if it did, then "reproductive isolation," which is a major tenet of speciation, could not occur. According to the evolutionary model, all species arose from a single-celled, asexual organism. Asexually reproducing organisms like bacteria reproduce by making a copy, or clone, of their self. How can we get sexually reproducing organisms from this? Asexual reproduction is a much more efficient method, and if evolution were true, then the "fittest" organisms would reproduce this way. The truth is that "male and female, he created them" (Genesis 1:27), which flies in the face of evolutionary dogma.

Macroevolution

In the evolutionary view, natural selection leads to speciation, which leads to macroevolution. Starr and Taggart define macroevolution as "large-scale patterns, trends, and rates of change among higher taxa"[9]. This definition describes a very predictable, mathematically based theory. The words and phrases used such as *patterns, trends*, and *rates of change* are all words used to describe mathematical relationships. Remember one of Stephen Hawking's definitions of a "good" theory is that it predicts outcomes. Therefore, we should expect macroevolution to be a very predictable, mathematically based theory. We should expect, like in Newton's Second law, Boyle's law, Charle's law, etc, an equation or equations that we can use to predict outcomes of future events.

> A good scientific theory should accurately predict outcomes, but macroevolution cannot do this.

However, we have no equations. What we do have is *evidences* of macroevolution, but since we have no direct observation, and no equations, we have no way to accurately predict the outcomes. A good scientific theory should accurately predict outcomes, but macroevolution cannot do this. Nevertheless, let's look at some of the common evidences for macroevolution, including the fossil record, the geologic record, radiometric dating of rocks, and morphological comparisons of existing species. I will discuss the strengths and weakness of each.

The fossil record

The fossil record is the preserved history of life on earth found in layers of sedimentary rock. Unless organisms are buried quickly, their soft parts decompose or scavengers make short work of them[10]. Conditions favor preservation when burial is rapid and lacking in oxygen[11]. This makes perfect sense, but now compare it to the following statements from the same book: "Stratified (stacked) layers of sedimentary rock are rich sources of fossils. They formed long ago by a gradual deposition of volcanic ash, silt, and other materials"[12]. And when discussing a picture of a fossil fish, "millions of years ago a bony fish died, and sediments gradually buried it"[13] Why are there such contradictory statements in the same book? How can you rapidly, yet gradually bury something (Fig. 6.2)?

The confusion lies in the clash of two beliefs about how the fossil record came about. Before Darwin, most scientists believed in the theory of *catastrophism*, taken from a literal interpretation of the events surrounding the flood described in Genesis chapter 6. These scientists, like me, believed most fossils formed during this flood. They also believed the earth was about 6,000 years old.

Darwin, in order to explain evolution, needed a different explanation involving longer times, and applied the then-developing theory of *uniformity*. The new theory hypothesized that layers of sedimentary rock formed gradually over millions of years, rather than the thousands proposed in the Bible. The confusion in today's secular biology textbooks comes from trying to justify uniformity amid the overwhelming evidence for catastrophism. We will discuss a major example momentarily.

If evolution were true, then there should be billions and billions of transitional fossils. The fossil record should contain multitudes of evolutionary transitions, including all the "failures." For example, if bats evolved from a mouse-like animal, then there should be untold numbers of fossil mouse-bats, some with long tails and stubby wings, others with enlarged forearms but underdeveloped wings, etc. However not one single mouse-bat that has ever been found, much less any cow-whales, feathered lizards, frog-fish or ape-men.

Darwin considered the lack of transitional forms "the most obvious and gravest objection which can be urged against my theory"[14]. Today, almost 150 years after Darwin said this, we still don't have any legitimate transitional forms. Colin Patterson, once a senior paleontologist at the British Museum of Natural History, which houses a collection of over 60 million fossils, said "If I knew of any [evolutionary transitions], fossil or living, I would certainly have included them [in my book *Evolution*][15]"

If anyone had seen a transitional fossil, I think it would be Professor Patterson! We have uncovered millions and millions of fossils since Darwin wrote *Origins*. As Dr. Duane Gish of the Institute for Creation Research likes to say, there should be "billions times billions" of transitional forms in the fossil record. Yet we have *none*.

Figure 6.2. The myth of fossil creation by gradual burial. Top: On a distant lakeshore, a bass lies dead. Middle: Picking up the scent trail, a curious, yet hungry feline approaches. Bottom: Scavengers make sure that gradual burial rarely takes place.

The geologic record.

Cambrian Explosion: Scientists developed a geologic timescale to estimate the age of fossils recovered from the earth's crust. Scientists hypothesized that the oldest fossils would be at the bottom layers of the earth's sedimentary rock strata, which makes sense from an evolutionary view. Names have been assigned to the various layers, such as Permian, Jurassic, etc. Scientists use radiometric dating, which we will discuss next, to assign ages to the different layers.

The first layer containing large quantities of animal fossils is right before the Cambrian era, and is referred to as the Cambrian explosion. It is an interesting layer, because there is a transition from no organisms to unbelievable numbers of organisms. Surprisingly, Starr and Taggart have one sentence on the subject on page 308, which says "all modern animal phyla emerged in adaptive radiations (the Cambrian explosion)"[16].

The Cambrian explosion is a major problem for the theory of macroevolution, but the problem is not discussed, which is a huge suppression of truth found in almost all secular biology textbooks[17]. Why is it not discussed? Well, if all modern animal phyla exist in this one layer, it doesn't suggest evolution over long periods, which is "supposed" to be what happened. What it does suggest is creation followed by a major flood with subsequent burial of large amounts of organisms.

A better name for the Cambrian layer might be "*the beach*." Think about it: with a flood of the magnitude described in Genesis 6-9, the first thing buried would be animals living in the sediments. The beach, with all of its seashells would get buried first, too. Understandably, any example of fossils found in the Cambrian layer will contain many shells.

Grand Canyon: Scientists use the theory of uniformity to propose old ages for the earth. Sedimentary rock layers are used as evidence, with each layer supposedly taking a year or more to form. A common example is the Grand Canyon, shown as Figure 19.7 in Starr and Taggart. The caption beneath the figure states "erosive forces of rivers carved the canyon walls"[18].

Think about the Grand Canyon, through which the Colorado River flows, and then think about a larger river, like the Mississippi. The Mississippi River has a much higher flow rate through it, yet there are no mile-deep canyons around it. Its banks do erode, but not rapidly, and they are usually not solid rock. So how did a little river like the Colorado carve a canyon thousands of feet deep in places, but also dozens of miles wide? Well, new evidence suggests massive glacial lakes once existed upstream of the Grand Canyon, and when the dams on the lakes broke, the massive rush of water eroded the layers away[19].

Figure 6.3. Canyon wall on American Creek, Katmai National Park, Alaska. Canyons never form gradually over millions of years, they form rapidly during major catastrophic events.

Examples of catastrophism causing major geologic changes abound, and the video "Mount St. Helens, Explosive Evidence for Catastrophe" by the Institute for Creation Research is an excellent source of examples of recent canyon formation and rapid sedimentary layer deposition.

Sea salt: Another argument from geology Darwin used was the saltiness of the sea. He believed seas started fresh and became saltier over time as rivers dumped salts into them. As evidence, he used the fact that all rivers contain at least a little salt, which they are constantly dumping into the sea. The evidence fits nicely with the uniformity model for geologic processes.

Matthew Maury, a contemporary of Darwin and considered as the "founder of oceanography," disagreed with this idea. Maury calculated that the amount of salt necessary to fill the seas would cover an area of 10 million square miles to a depth of 1mile. This seems like an impossible amount of salt to enter the sea by rivers or any other means. Maury said

> *"I once thought with Darwin that the sea derived its salts originally from the washings of the rains and rivers. I now question that opinion; for, in the course of the researches connected with*

the Wind and Current Charts, I have found evidence, from the sea and the Bible, which seems to cast doubt upon it. The account given in the first chapter of Genesis, and that contained in the hieroglyphics which are traced by the hand of Nature on the geological column as to the order of creation, are marvelously accordant; and they contain no evidence that the sea was ever fresh"[20].

Evidences for catastrophic change and for the sea always being salty are not discussed in secular biology texts, which is yet another suppression of information. To discuss such topics in a secular biology textbook would greatly weaken the evidence in favor of the principle of uniformity. Keep in mind though that all of this is evidence. No one was there to see it happen.

Radiometric dating of rocks.

This is a method used for measuring the age of rocks by studying the ratios of the isotopes of various elements found in them. An isotope has a different number of neutrons when compared to the most abundant form of a particular element. For example, most carbon atoms have 6 neutrons, but small percentages have 7, and an even smaller percentage have 8.

Most elements have at least one isotope, and some of these are radioactive. Radioactive elements are by nature unstable, and over time they actually lose parts of their nucleus and turn into a different element. By measuring the ratio of the radioactive element to its stable decay product, scientists can estimate the ages of rocks, and if they know the age of the rocks, they can estimate the age of fossils found near them. To use radiometric dating, the following assumptions must be made.

a. The isotope has decayed (turned into its stable product) at a constant rate.

b. The original amount of isotope must be guessed at.

c. The constant decay rate followed an exponential pattern.

We will discuss radiometric dating in more detail in Chapter 7, but for now just know that dates generated by this method are unreliable because of the impossibility of knowing whether these three assumptions are valid.

Morphological comparisons:

Skeletal comparisons: Morphological comparisons are comparisons of the shapes and structures of different parts. In the evolutionary view, common design equates with common ancestry. Examples abound in Starr and Taggart of comparative morphology. For example, Figure 19.10 compares the forearm bones of a fossil reptile to several other animals, including a bat, a penguin, and a human, and suggests this is evidence that they all evolved from this reptile[21]. There is no denying the similarities of structure between different species. However, if these animals were really ancestors of this "stem reptile," then there should be billions and billions of transitional fossils.

Haeckel's embryo drawings: Another common example of comparative morphology is with the early stages of embryos. If evolution is true, then supposedly early stages of embryos should be similar. Figure 19.13 of Starr and Taggart compares early embryos of a fish, reptile, bird, and mammal[22]. However, according to the Discovery Institute, research has shown that scientists based the comparison on faked sketches by a contemporary of Darwin, Ernst Haeckel. In reality, early stage embryos of different species are quite different[23]. A good resource discussing the faked embryo drawings is the video "Icons of Evolution."

Human/chimp DNA comparisons: Evolutionists consider chimpanzees as man's common ancestors. Figure 19.14 in Starr and Taggart shows some comparisons in skull shapes, and the text states that "More than 98 percent of human DNA is identical with chimpanzee DNA"[24]. It doesn't take a PhD in biology to see the similarities between chimps and humans. However, think back to our studies of genetics, mutations, and probability. If we have 98% of our DNA in common with chimps, what about the other 2%?

> If we have 98% of our DNA in common with chimps, what about the other 2%?

Two percent sounds like such a small amount, until you remember that our genome equates to about 3.2 billion base pairs of DNA. Two percent of 3.2 billion is still 64 million, which doesn't sound so small anymore. On top of that, how do you get 64 million base pairs to organize themselves spontaneously to make new beneficial genes, when every study concludes that most mutations are harmful? The answer is, you can't.

Morphological comparisons can help us learn about similarities that exist between species. However, there are two ways to interpret the results. The evolutionist will say the similarities point to a common ancestor, while the creationist will say the similarities point to a common Designer. We can interpret the same set of data two completely different ways because predicting the past by using morphological comparisons is not real science, but is speculation.

Summary

Let's apply deductive reasoning and the three classes of truth to summarize what we have just learned. Recall that deductive reasoning is about applying rules. We call rules our "major premises," and we apply them to discover new truths. Some examples of major premises include God's word, Euclid's postulates, and standards for weights and measures. We know that even if our major premise is false, we can still have a logical argument, but it will lead to one or more false conclusions.

When trying to answer the question "where did all the species come from?", three possible major premises are 1) God made them in 6 days 2) God made them, and evolution was the "second cause" 3) God didn't make them and they evolved from nothing. Most Christians believe major premise 1 or 2. Darwin's major premise was number 3, and the whole purpose of his writing *Origin of Species* was to "show that species had not been separately created"[25]. Think about this though. How can you have three major premises that are true? You can make logical arguments based on all three major premises, but just because an argument is logical doesn't mean it's true.

Why is it there are so many major premises for answering the question "where did all the species come from?" If someone asks "how many minutes are in an hour?", the only correct answer is 60. When someone asks the question "How many quarts are in a gallon?", the only correct answer is 4. If someone were to give a different answer for either question, anyone who knew the truth would immediately correct their mistake.

The question of "where did all the species come from?" has only one correct answer, and the confusion comes because of man's sin nature, because we suppress His truth and exchange it for the lie (Romans 1:18-25). In Chapter 1 we discussed Leonhard Euler's most popular book "*Letters to a German Princess*" wherein he divides truth into three categories: truths of the senses, truths of understanding, and truths of belief. We can try to use our senses to answer the question of origins, but we will always come up short because we were not there to see it happen. We can use our reasoning to answer the question of origins, but to do that we must have a major premise, and if our major premise is false, then everything based on that major premise is false.

> The question of "where did all the species come from?" has only one correct answer.

Ultimately, whatever you believe as truth about origins is based on faith. No one observed Creation's beginning, ruling out science and man's reasoning alone as ways to answer this question. However, there is a historical truth that does reveal the answer to this question, and that is the Bible. I was not alive during World War 2, yet I believed it happened. I have read about it, and I have talked to someone that was there. There are also many truths of the senses and truths of understanding that support my belief that World War 2 happened. In the same way, I was not alive when the earth was formed and life was created, yet I believe it happened. I have read about it in the Bible, and I have talked to the One who was there through prayer. There are many truths of the senses and truths of understanding that support my belief that God created the earth and all the species in 6 days.

While healthy in his youth, Darwin led an invalid existence for the remaining 40 years of his life. While his medical advisers never reached definite conclusions as to the cause of his many illnesses, recent emphasis has been towards neurotic or psychotic causes.[26] It was during this time that he wrote "*On the Origin of Species*" and "*Descent of Man*," as well as his autobiography, in which he referred to Christianity as a "damnable doctrine."[27] Richard Dawkins, a famous defender of the evolutionary view, stated in a 1989 editorial in *The New York Times* that "It is absolutely safe to say that if you meet somebody who claims not to believe in evolution, that person is ignorant, stupid or insane (or wicked, but I'd rather not consider that)."

From what I have read about Darwin, he was a good scientist. However, his complete lack of knowledge of genetics and molecular biology, poor math skills ("I have deeply regretted that I did not proceed far enough at least to understand something of the great leading principles of mathematics")[28], and possible insanity make it difficult to trust his works.

Now, I am not trying to "slam" Charles Darwin here. The Bible says in Romans 3:10 that "none are righteous," and in 3:23 that "all have sinned and fall short of the glory of God," so that includes you, me, Darwin, and everyone else. What I am trying to show here is that everywhere you look in evolution, from the theory itself to those who write about it, suppressions and exchanges of Truth abound.

There was a time when Darwin did not "in the least doubt the strict and literal truth of every word in the Bible"[29]. However, disbelief crept over him at "a very slow rate, but was at last complete"[30]. Hebrews chapter 6 has a vivid description of what happens to people who "have tasted the good word of God" and then "fall away," and it ain't pretty! Those having tasted the good word of God will either "bear herbs useful to those by whom it is cultivated" and receive a blessing from

> You can either believe that the Bible is the correct account of the origin of species, or you can believe something else happened.

God, or they will "bear thorns and briers" and their end "is to be burned."

So what about you? What do you believe? Ultimately, faith directs what you believe as truth about the origin of species, and you can either believe that the Bible is the correct account of the origin of species, or you can believe something else happened.

The Bible says that faith in God's word is a gift from God. Ephesians 2:8-9 reminds us there is nothing we can do to earn this gift of faith. It is something only God can give us, we cannot earn it. God's word reveals there is more to this life than what is happening on earth right now. One day, those who do His commandments will "enter through the gates into the city," while those on the outside will include "whoever loves and practices a lie" (Revelation 22:14-15).

7 IT'S NOT THAT OLD
Origins and the Age of the Earth

When Charles Darwin wrote "*Origin of Species*," he was trying to answer the question "Where did all the species come from?", and we covered this topic in Chapters 5 and 6. Darwin did not propose a theory about the origins of the *chemical and physical world, and of life itself,* other than to speculate that it possibly happened in a "warm little pond"[1]. One point I made in Chapter 6 is that *evidence* about our past is not real science, but is speculation. Science is *not* about "truths of belief," but it *is* about "truths of the senses." All the evidences supporting macroevolution are likewise evidences supporting special creation by God because they are truths of belief. We weren't there to watch their formation.

Real science *generates conclusions* describing a collection of observations, or facts. Speculation on origins works in reverse, and tries to *generate facts* based on the conclusions we make regarding the past. In Chapter 7, we will learn what secular biology textbooks are teaching students about the origins of the chemical and physical world, and of life itself, and we will discover some more suppressions and exchanges of truth. We will study the big bang theory, the Miller-Urey experiment and chemical evolution, and will revisit methods used to guess at the age of the Earth and look at evidences for a young earth.

The Big Bang Theory

Everything is made of chemicals. Currently scientists know of over 100 elements, and these elements combine in different ways to form all the living and nonliving matter making up the universe. But where did all of these chemicals come from? The prevailing view taught in secular biology textbooks is that all chemicals had their origins in a phenomenon known as the "big bang." This is how scientists describe the Big Bang:

> [About 12-15 billion years ago] all of space and time were compressed into a hot, dense volume about the size of the sun. That incredibly hot, dense state lasted only for an instant. What happened next is known as the **big bang**, a stupendous, nearly instantaneous distribution of matter and energy throughout the universe. About a minute later, temperatures dropped a billion degrees. Fusion reactions created most of the light elements, including helium, which are still the most abundant elements in the universe. Radio telescopes can detect cooled, diluted background radiation-a relic of the big bang-left over from the beginning of time.[2]

The big bang theory suggests that galaxies, stars, and eventually our solar system formed because of this explosion, and many scientists believe it accurately explains how the universe formed. However, it is not a truth of the senses because we weren't there to see it happen! Amazingly, many scientists have no problem believing the universe formed "almost instantaneously," but they have a huge problem with God's word saying He created the universe, the earth and all living things in 6 days! Six days is an "almost infinite" amount of time compared to the "almost instantaneous" formation of the universe.

> Amazingly, many scientists have no problem believing the universe formed "almost instantaneously," but they have a huge problem with God's word saying He created the universe, the earth and all living things in 6 days!

Now, if God wanted to use a "big bang" to start things off, then that's fine with me. In fact, a "big bang" fits well into the evidence for a 6-day creation.

Just like the evidences for macroevolution, the big bang theory can be used as evidence for either evolution or Creation. Of course, a secular textbook will not use it as evidence for creation or a Creator, even though anyone who thinks critically about the theory will most likely ask "where did the stuff to make the big bang come from?" Likewise, you will never find question "how was life *created* from chemicals?" in a secular biology textbook, but the question "how did life *evolve* from chemicals?" will be, with a lengthy explanation to follow.

Chemical evolution

As Starr and Taggart describe it, *life* itself is merely a "continuation of the physical and chemical evolution of the universe"[3]. So how do secular biology textbooks say life originated from nonliving chemicals? Well first, to explain the evolutionary transition from chemicals to life logically, the evolution of organic chemicals must be discussed. Living things make *organic chemicals*, and one particularly important group of organic chemicals is amino acids. Amino acids are found in DNA, and amino acids are the building blocks of protein molecules. If evolution were true, these molecules had to evolve first.

Most secular biology texts use the Miller-Urey experiment to help explain chemical evolution. The following is a summary of a description of the Miller-Urey experiment, written by the Discovery Institute in their 2003 report addressed to the Texas State Board of Education[4]:

> *In the early 1950's, many scientists believed that the earth's atmosphere was composed of water vapor, hydrogen and hydrogen-rich gases such as methane and ammonia. Stanley Miller, a graduate student in the laboratory of University of Chicago professor Harold Urey, mixed these gases in a glass chamber and applied a spark to simulate lightning. A week later, he found that he made a mixture that contained some organic chemicals, including amino acids. He reported his results in 1953, and the Miller-Urey experiment has since been incorporated into many biology textbooks as the starting point for origin-of-life research. One major problem with using the Miller-Urey experiment in current secular biology textbooks is that by 1980, most scientists had concluded that the Miller-Urey experiment was irrelevant. What happened was that in the 1960's, scientists began to uncover evidence that the early earth's atmosphere was primarily carbon dioxide, nitrogen and water vapor, and when these*

chemicals were tested using the Miller-Urey apparatus, no amino acids were produced.

Starr and Taggart describe the Miller-Urey experiment on page 329 of their text[5]. The Discovery Institute gave the textbook a grade of "F" for its description. They concluded the textbook "fails to inform students that the Earth's early atmosphere was probably quite different from the mixture of gases used in the experiment, or that when the experiment is repeated with a realistic mixture it does not work. Under realistic conditions, the molecules produced include toxic chemicals such as cyanide and formaldehyde, but not amino acids." In addition "the text completely omits any discussion of problems with the experiment, even though those problems have been widely reported in the scientific literature for several decades."[6]

Figure 7.1 Mixing chemicals in an attempt to prove evolution,
as done in the Miller-Urey experiment, is not real science, but is speculation.

This amazing suppression of truth existed in all biology textbooks involved in the 2003 Texas textbook adoption. All the books either received a "D" or an "F" from the Discovery Institute for their description of the Miller-Urey experiment[7].

The Discovery Institute says "the truth is that scientists are as far as ever from understanding how life's building blocks formed on the early Earth, and even farther from understanding how cells formed from such building blocks"[8]. So why is Miller-Urey still in secular biology textbooks? The reasons are many, including, laziness, fear, and just plain "loving the lie" (Revelations 22:14-15). Those promoting an evolutionary worldview have no better explanation, and to convince everyone that evolution happened, they must use lies to do it. I may sound like a broken record here, but it is important to keep in mind that all "scientific evidence" regarding origins is just speculation, no matter if it was Miller and Urey, or the scientists who disproved them. Since discussions of origins are not based on real scientific data, they are not really topics for biology textbooks. Philosophy maybe, but not biology.

Determining the age of the Earth, evidences for a young earth.

One way evolutionists try to tie together physical, chemical and biological evolution is by using experimental methods to measure the age of the Universe, the Solar System, and the Earth. Supposedly, the Sun formed around 4.6 billion years ago, followed by the Earth and the rest of the solar system. Life on earth supposedly formed about 3.8 billion years ago.

The main method scientists use to determine the age of the earth is a method called *radiometric dating*. This method, discussed briefly in Chapter 6, measures the age of rocks by studying ratios of isotopes of various elements found in them. Most elements have at least one isotope, and some of these are radioactive. Radioactive elements are by nature unstable, and over time they actually lose parts of their nuclei and turn into a different element. By measuring the ratio of the radioactive element to its stable decay product, scientists can estimate the ages of rocks. To use radiometric dating, the following assumptions must be made.

1. The isotope has decayed (turned into its stable product) at a constant rate.
2. The original amount of isotope must be guessed at.
3. The constant decay rate followed an exponential pattern.

To understand radioactive dating, I will attempt to explain it using an analogy of speed. If you drove 100 miles at an average speed of 50 miles per hour, you can easily calculate that it took two hours to drive this distance. Now, in a different situation, let's say you were standing by a highway and you saw a car go by, and you wondered "how long has that car been driving"? Well if you knew the speed limit on the highway, you could assume the car was driving at the speed limit, but you still had no idea where the car started from, so you guessed at the starting point. Since you guessed the starting point and you assumed the car was driving the speed limit, your estimate of how long the car had been driving would not be very reliable.

Now, apply this analogy to determining the age of a rock. Let's say you found a rock and wondered "how long ago was this rock formed?" Well, if you knew the speed, or rate at which radioactive isotopes within the rock decayed, and you guessed a starting point, or initial amount, and you assumed the element decayed in a predictable pattern, then you could estimate the age of the rock, but the estimate would not be very reliable.

In his book "*That Their Words May Be Used Against Them*," Henry Morris quotes dozens of scientific articles discussing the unreliability and unpredictability of the radiometric dating method[9]. This is in contrast to Starr and Taggart, which says the method has an "error factor of less than 10%"[10].

If that doesn't surprise you, consider this: scientists use radiometric dating to determine the age of fossils, even though most fossils lie in sedimentary rock, while radiometric dating only works for volcanic rock or ashes. Therefore, scientists must find volcanic rock in the same relative position as the fossils they are studying, and therefore must assume the volcanic rocks formed at the same time the sedimentary rocks formed! Because of the impossibility of knowing whether the assumptions made to use this method are valid, radiometric dating is very unreliable.

Evidences for a young earth.

As long as people speculate about the age of the earth in order to fit an evolutionary framework, others should make counter speculations fitting a Biblical framework. However, many prominent Christian scientists, ones who have publicly rejected Darwinism, still believe the earth is billions of years old. Hugh Ross, for example, describes the creation days as "geologic ages," and has created a chart correlating the two (www.reasons.org)[11]. The chart seems logical, but a logical statement is false if the major premise is false, and the major premise for this "day-age" theory is that God didn't mean a 24-hour day.

Galileo once said that "the Bible tells us how to go to heaven, not how the heavens go."[12] He was saying the Bible is not a science textbook, and I agree. However, Galileo also said that "Both the Holy Scriptures and Nature proceed from the Divine Word" and that "the two truths can never contradict each other"[13]. Therefore, when the Bible speaks of observable phenomena, we better listen, and if our scientific research doesn't match up with Scripture, then it is man who has erred.

A close look at the Bible reveals that God created for 6 days, and the earth is somewhere around 6 thousand years old. A variety of evidences support this age, all based on rate-related measurements. The Institute for Creation Research performed one of the most recent experiments sup-

> A close look at the Bible reveals that God created for 6 days, and that the earth is somewhere around 6 thousand years old.

porting a recent creation[14]. They measured helium diffusion rates in zircons (minerals composed of zirconium and silicon) drilled from deep in the earth and found the rates correlated well with a model developed to predict helium diffusion rates for a young-earth model. The measured rates correlated poorly with an old earth model. Their results also provided strong evidence for accelerated decay of radioisotopes in early earth history. Recall that one of the major assumptions used in radiometric dating is that radioisotope decay rates have been constant throughout earth's history. ICR's research provides evidence to the contrary.

Another experiment by ICR providing evidence for a young earth involves analysis of *radiohalos*. Radiohalos are minute spherical zones of discoloration surrounding tiny mineral crystals imbedded in larger mineral grains such as granite[15]. The radiohalos are like scars, resulting from damage inflicted to the surrounding crystal by radioactive emissions.

Of most interest are the radiohalos produced by polonium (Po). The three radioactive Po isotopes have very short half lives (the time for half of the material to decay into another element), from microseconds to days. The fact that radiohalos from Uranium and Thorium, which supposedly take 100 million years to form, are found along with the Po halos, is evidence the materials rapidly cooled from a liquid to a solid state (radiohalos require a solid to form). Current theories suggest that rocks containing these halos required hundreds of millions of years to cool and solidify.

Evidence of Ur and Th halos alongside the shorter-lived Po halos is further evidence for accelerated decay rates in the past. If it took hundreds of millions of years for these rocks to cool and solidify, then all Po should have decayed before solidification, leaving only Ur and Th halos. Since all current radiometric dating methods assume the decay rate of radioisotopes is constant over time, this contradictory evidence by researchers at ICR provides plenty of evidence in favor of a young-earth model.

Summary

The field of science is changing rapidly. We are constantly learning that many ideas assumed to be true are quite the opposite. As an example, from 1924 to 1988 the visitor's sign above Carlsbad Caverns in New Mexico pronounced the cave system was at least 260 million years old. In 1988, the sign was changed to read 7 to 10 million years old. Then, for a little while, the sign read that the caverns were 2 million years old. Now, the sign is gone[16]. Wollemi Pine trees, thought to be extinct for 150 million years, were discovered alive and "unevolved" in Australia in 1994[17].

If you look for it, plenty of evidence exists revealing that current methods for dating fossils, the earth, and the universe are erroneous and misleading. All the methods, whether they predict an old or young earth, are speculations at best because we were not there to make the direct observations. Did God use a big bang to create everything? How did he make the chemicals required by living organisms? Exactly how old is the earth? I am completely satisfied with answering all of these questions as "I don't know."

> Plenty of evidence exists revealing that current methods for dating fossils, the earth, and the universe are erroneous and misleading.

What I do know is that I could speculate all day long as to what the answers to those questions are, and I could even do some scientific experiments to support my answers. However, my purpose in writing this book is not to speculate, but to focus on the certainty of God's word. He said He created everything in 6 days, and if you read the Bible closely, you will find the earth is around 6,000 years old.

In Starr and Taggart's book, Chapter 20 is titled "The Origin and Evolution of Life"[18]. As we have seen, this chapter is full of suppressions of truth, but it starts out incredibly well. In fact, it starts out with a phrase from Genesis, "In the Beginning ...," stamped above the introductory paragraph. However, as could only happen with a secular text, the Truth is exchanged for the lie. What the students who use this book, and all of us, desperately need to hear though is the Truth. We need to hear that "*In the beginning,* God created the heavens and the earth" and

> "*In the beginning was the Word, and the Word was with God, and the Word was God. He was in the beginning with God. All things were made through Him, and without Him nothings was made that was made. In Him was life, and the life was the light of men. And the light shines in the darkness, and the darkness did not comprehend it.*"(John 1:1-5).

8 TRUTH
Biblical Creationism

One secular biology textbook discusses the following erroneous account of a controversy involving the famous scientist Galileo in the early 1600's:

> *Nicolaus Copernicus studied the planets and concluded the Earth circles the sun. Today this seems obvious enough. Back then, it was heresy. The prevailing belief was that the Creator made the Earth-and, by extension, humans-the immovable center of the universe. Later a respected scholar, Galileo Galilei, studied the Copernican model of the solar system, thought it was a good one, and said so. He was forced to retract his statement publicly, on his knees, and put the Earth back as the fixed center of things. (Word has it that when he stood up he muttered, "Even so, it does move.") Later still, Darwin's theory of evolution ran up against the same prevailing belief.*[1]

This paragraph makes it sound like Galileo's principal conflict was with the Creator, and therefore a religious controversy, when it was actually with the Aristotelian scientists of his day[2]. In fact Galileo's personal philosophy paralleled that of Francis Bacon, as he said "God is known ... by Nature in His works, and by doctrine in His revealed word"[3]. It is interesting to read about the Scriptures that supposedly support a fixed, nonmoving earth (Joshua 10:12-13, Ecclesiastes 1:5, Isaiah 38:8). No mention is made in these Scriptures that the earth is motionless, only that the sun "moves," so it was assumed that since the sun moves, the earth doesn't. Philosophers of Galileo's day made this assumption not as an attempt to reconcile Scripture with direct observation, but as an attempt to reconcile Scripture with man's reasoning in the form of Aristotelian philosophy. In reality, Galileo's heart was to show that "The Holy Bible can never speak untruth"[4], while Darwin's heart was to "show that the species had not been separately created"[5]. Galileo used direct observations to support his theories, while Darwin used indirect evidence and the faked data of Thomas Malthus to support his theory. This paragraph in Starr and Taggart is an attempt to show that science and the one true religion, Christianity, don't mix, when in reality the two are inseparable. It is another unfortunate exchange of truth found in secular biology textbooks.

How has the story become so twisted? I think Scripture has the answer, and it lies in the truth that "*none are righteous*" (Romans 3:10). One thing unrighteous people do is "*exchange the truth of God for the lie*" (Romans 1:25), and in order to exchange the Truth, they have to use lies and "evidence," but no direct observations.

So what about the Biblical account of creation? How do we know that it is true? We concluded Chapter 7 by saying that faith directs what you believe as truth about origins. You can either have faith that "*the worlds were framed by the word of God, so that the things which are seen were not made of things which are visible*" (Hebrews 11:3), or you can believe something else happened. Either the Bible is the greatest lie of them all, or it is true.

God has given me faith to believe His word, and I hope you also have received God's gift of faith, otherwise what we are about to discuss will seem like foolishness. In this chapter we will look at a Biblical account of the past, present, and future of the Earth. We will also discuss attempts to reconcile the Bible and evolutionary frameworks. A valuable resource when writing Chapter 8 was *The Defender's Study Bible*, published by the Institute for Creation Research[6].

A Biblical account of the history of the world

The world that was then

The Bible says it took six days to create the earth and all life. I believe these were 24-hour days, based on the discussion of morning and evening in Genesis chapter 1 and 2, as well as the comparison of our work week with God's creation week in the 4th commandment (Exodus 20:11). There are many times in Scripture when God uses symbolism and parables to relay His message using an abstract truth. Recall that an abstract truth applies to a variety of situations. There are also specific truths, such as the Ten Commandments. "Do not steal," "Do not murder," "Rest on the 7th day." Exodus 20:11 is part of the Ten Commandments, and is proof enough that God created the Earth and all life in 6, 24-hour days.

Some say God was referring to "periods of time," and just used the word "day" as an analogy. After all, the Hebrew translation of the word, as well as the Webster's collegiate dictionary gives one possible definition of day as "period of time." If that is true, then God could have created the earth in 6 seconds, because 6 seconds is also a "period of time"! The amount of time taken to create everything we see today will be argued forever because no one was there to see it happen, and there is no way we can provide convincing evidence through real scientific observations. I trust that when God said 6 days, he meant 6, 24 hour days, just like when Jesus said "and the third day He will rise again (Luke 18:33)," he meant 3, 24 hour days, not 3 unknowable "periods of time."

Some other major points of "the world that was then," include a global greenhouse in the form of a vapor canopy (Genesis 1:6, 2:5-6). God gave man the task of taking care of His creation (Genesis 1:28), it was cursed for man's sake (Genesis 3:17-19) corrupted by man's sin (Genesis 4:11, 6:11-12), and then destroyed by water (Genesis 6-8).

Figure 8.1. God created everything in six days, and on the seventh, rested from His work.

The world that is now

The current creation is held together by its Creator (Colossians 1:17). An amazing evidence of this truth exists inside the nucleus of every atom. An atom's nucleus contains protons and neutrons. Protons have a positive charge, and the scientific law of charges says that like charges should repel one another. How is it then that the protons, packed together in the nucleus of an atom, are not flying away from one another? This phenomenon is explained away by using words like "strong nuclear forces," which scientists do not know much about.

The world that shall be

Upon Christ's return, suffering and death will no longer exist (Revelation 21:4). Everything will be made new again (Acts 3:21; Revelation 21:5) and it will be eternal (Daniel 12:3; Isaiah 66:22). Sin will no longer exist either (2 Peter 3:13; I John 3:1-3). What an amazing time that will be!

Conclusion:

In Chapter 7, we looked at many evidences used to prove evolution. Most of those evidences, such as the fossil record and the geologic record, are also used to prove creation by God. How can we use the same observations to prove two totally different things? The answer is that evidences prove neither evolution nor creation, because no one was there to see either one. There is nothing wrong with studying geologic formations or fossil beds, but we must be careful about making useful conclusions regarding our observations and not just speculations.

Trying to use the scientific method to prove what happened in the past is not real science but is speculation. It is guessing in other words. In Chapter 7 we defined a good scientific theory as one that accurately describes a large class of observations and makes definite predictions about the results of future observations. We can say that "this experiment suggests the earth is 6,000 years old," or "fossils collected from this site suggest clams were here before fishes." However, we cannot make a good scientific theory from this data because it is useless for predicting future observations. When studying the past, we must be careful to separate science from speculation. At the same time though, we must acknowledge God's command in Proverbs 26:5 and "answer a fool according to his folly, lest he be wise in his own eyes." With this in mind, let's look at some speculation that attempts to reconcile the Bible within an evolutionary framework.

Attempts to reconcile Biblical and Evolutionary frameworks

Literary Framework Theory:

The literary framework theory takes the evolutionary ages as literally true. This is also known as Theistic Evolution. A theistic evolutionist believes that God used evolution as His mechanism for creating life. Many scientists at Christian colleges believe this. We have already presented many of the untruths required to discuss evolutionary theory, so a Christian who believes the literary framework theory is deceived. Colossians 2:8 is a warning to Christians against such deception.

Day-Age Theory:

This theory assumes the evolutionary ages to be the same as the Creation week. For example, Day 1 is a certain evolutionary age, Day 2 refers to another, etc. This is also known as "Progressive Creationism," which adheres to a belief that intermittent acts of creation were interspersed between long ages of either slow or no evolution. Hugh Ross is the major advocate of Progressive Creationism (www.reasons.org). This theory sounds logical, but speculates that God didn't mean 24-hour days during the creation week. I recently read a book by a sincere Christian and incredibly gifted scientist named Henry S. Schaefer. In his book[7], he makes it clear that he believes Ross's argument. However, I could not find anywhere where Schaefer mentions Exodus 20:11. It is impossible to interpret Exodus 20:11 any other way than a 7-day week like ours, and to believe otherwise is to be deceived.

For whatever reason, Dr. Schaefer does not discuss Exodus 20:11 in his book, *Science and Christianity,* but he should. However, discrediting everything Dr. Schaefer has to say because of this one error would be like discrediting everything King David said because of his sins.

Gap theory:

The gap theory suggests completion of the evolutionary ages before the Creation week (Genesis 1:1&2), with the ages terminated by a global catastrophe followed by reconstruction. The theory finds a place to "fit" evolution into the Biblical account. The fallacy of evolution in the first place, as well as the extreme speculation used in developing this theory, makes it worthless at best.

Conclusion:

We cannot know exactly how God created life. It is an incredible mystery, and all we can do is speculate. God wants to reveal himself to us through His creation (Romans 1:20), but at the same time man should never think he can "find out the work God does from beginning to end" (Eccl. 3:11). As a result, some of our speculations are more illogical than others, especially when they attempt to reconcile evolution with the Bible. God is not lying to us, so if he says he made the earth in 6 days, why not just believe it?

> God is not lying to us, so if he says he made the earth in 6 days, why not just believe it?

Just like it was wrong for the philosophers and church leaders of Galileo's day to use something the Bible did not say to prove that a man's (Aristotle) reasoning was right, it is wrong for us to use something the Bible did not say to prove a man's (Darwin) reasoning was right. The Bible doesn't say "6 minutes," "6 seconds," "6 million years," or "6 periods of time." It says "6 days." No one was there to observe creation, so it cannot be proven scientifically, but we can be sure that God is real and His word is true.

Questions from the Bible

God wants us to ask questions, as His word says for us to think about what we are doing (Proverbs 14:8). We have spent a lot of time asking questions, and we have seen that others have questions about God's word and are skeptical of it. I thought it would be good to end Chapter 8 by contemplating some questions regarding Creation found in the Bible. Job 38-41 is an amazing set of chapters where God is questioning Job, and is commonly referenced as evidence that dinosaurs and humans lived at the same time. Take some time now to read Job 38-41, and when finished, read the following passage from Isaiah:

> *Who has measured the waters in the hollow of His hand,*
> *Measured heaven with a span*
> *And calculated the dust of the earth in a measure?*
> *Weighed the mountains in scales*
> *And the hills in a balance?*
> *Who has directed the Spirit of the Lord,*
> *Or as His counselor has taught Him?*
> *With whom did He take counsel, and who instructed Him,*
> *And taught him the path of justice?*
> *Who taught Him knowledge,*
> *And showed Him the way of understanding?*
>
> *Behold, the nations are as a drop in a bucket,*
> *And are counted as the small dust on the scales;*
> *Look, He lifts up the isles as a very little thing.*
> *And Lebanon is not sufficient to burn,*
> *Nor its beasts sufficient for a burnt offering.*
> *All nations before Him are as nothing,*
> *And they are counted by Him less than nothing and worthless.*
>
> *To whom then will you liken God?*
> *Or what likeness will you compare to Him?*
> *The workman molds an image,*
> *The goldsmith overspreads it with gold,*
> *And the silversmith casts silver chains.*
> *Whoever is too impoverished for such a contribution*
> *Chooses a tree that will not rot;*
> *He seeks for himself a skillful workman*
> *To prepare a carved image that will not totter.*

Have you not known?
Have you not heard?
Has it not been told you from the beginning?
Have you not understood from the foundations of the earth?
It is He who sits above the circle of the earth,
And its inhabitants are like grasshoppers,
Who stretches out the heavens like a curtain,
And spreads them out like a tent to dwell in.
He brings the princes to nothing;
He makes the judges of the earth useless.

Scarcely shall they be planted,
Scarcely shall they be sown,
Scarcely shall their stock take root in the earth,
When He will also blow on them,
And they will wither,
And the whirlwind will take them away like stubble.

'To whom then will you liken Me
Or to whom shall I be equal?' says the Holy One.
Lift up your eyes on high,
And see who has created these things,
Who brings out their host by number;
He calls them all by name,
By the greatness of His might
And the strength of His power;
Not one is missing.
Isaiah 40:12-26

After contemplating the questions in these verses, obviously an atheist or agnostic like Charles Darwin would either have to ignore them or outright reject them. Such verses remind us of God's love, power and infinite wisdom, and they remind us too that we are not as smart as we think we are! Fortunately, most of the good, noteworthy scientists listened to Scriptures such as Isaiah 40.

Take for example Gerard Mercator, born in Europe in 1511. From his youth, geography was a primary object of his study[8]. His university teaching, however, left him ill-prepared for the mathematical skills required for mapmaking and surveying, so with the help of the mathematician Gemma Frisius, he educated himself at home in mathematics. The study of nature amazed Mercator[9], and geography was the most suitable branch of knowledge for explaining the structure and mystery of God's creation[10]. In 1537, he finished his first map, a map of the Holy Land. On it were written many Scriptures, including this passage from Matthew 17:5 "This is My beloved Son, in whom I am well pleased. Hear Him!"

Mercator's maps cried out to their readers to listen to the Scriptures[11]. Mercator's mapmaking skills affect practically every map made today, and the words "Mercator projection" are commonly seen on modern maps.

This Christian man of God sought the Lord, and questioned the traditions of his day, including the philosophies of Aristotle and some of the erroneous teachings of the Catholic Church. Mercator was almost beheaded because of his questioning and his following of the teachings of

> Today, dogmatic teaching of Aristotle has changed to dogmatic teaching of Darwin.

the Christian reformer Martin Luther. Today, dogmatic teaching of Aristotle has changed to dogmatic teaching of Darwin, and people still persecute Christians. We should not expect to live in a perfect world, and the verses in 2 Peter 3 are evidence enough of that. What we should expect is that until His kingdom comes, hollow and deceptive philosophies will exist, and people will continue looking for reasons to deny God.

I hope that you will follow the examples set by men like Galileo and Mercator, and question the "traditions of men." If their philosophies don't match up with what God's word says or His works reveal, then reject them. Jeremiah 33:3 says "call to me and I will teach you great and unsearchable things that you do not know." Look for His answer in His word and in His works.

9 MEET YOUR MAKER
Intelligent Design

While conducting research for Chapters 3 and 4, I ran across an interesting paragraph in the 3rd edition of *Human Molecular Genetics*[1]. It seems that the amount of product manufactured by genes is more important in some genes than others. If certain genes create too much or too little of a product because of a mutation, then abnormal phenotypes can occur, resulting in disease or death of the individual. Scientists refer to this property of mutated genes as *dosage sensitivity*. The two sentences that captured my attention were questions the authors asked. They said "One might reasonably ask why there should be dosage sensitivity for any gene product. Why has natural selection not managed things better?"[2] They go on to speculate that natural selection is working and that is why most genes are not dosage sensitive.

What really got my attention was the next sentence when they stated there are several genes in the human genome that *are* dosage sensitive, and they have very important functions such as determining the timing of development and metabolism. It surprised me to see the authors of this secular textbook describe something they could not explain with an argument from natural selection, which brought to mind the following question: If a textbook written from an evolutionary viewpoint admits that it cannot explain something within the framework of natural selection, then what explains it?

The only logical explanation is that *it was designed to work that way*, but as we have seen, the current versions of secular science textbooks do not always deal with ideas logically. For a textbook to say that something was designed might lead to a student asking "who was the designer?", which in turn may teach students that science and religion do mix, which is what current textbooks try to teach students is *not* the case.

Figure 9.1. Intelligently designed eagles. Imagine the absurdity of saying the bald eagle on the left was not designed by an intelligent being, while the F-15 Eagle on the right was.

Fortunately, as the fallacy of evolution becomes more and more exposed, a larger crowd of people, tired of the lies, will demand textbooks that state the obvious, such as the truth that life *was* designed. The theory currently at the forefront of a large movement in America is "intelligent design" (ID for short).

Just like any theory, ID has a major premise. According to Phillip Johnson, a leader in the ID movement, the major premise is "that intelligence is a real phenomenon which cannot be reduced to material causes and which can be identified scientifically"[3]. In Chapter 9, we will look at the history of design theory, learn about intelligent design theory and the current "ID movement," and conclude with a discussion of the theory's strengths and weaknesses.

History of design theory

Design theory is a view that nature shows tangible signs of having been designed by a preexisting intelligence[4]. Design theory has been around at least since ancient Greece, and was a commonly accepted theory before Darwin's theory of evolution. William Paley was the prominent design theorist in the 19th century, and in 1802 published a book titled *Natural Theology*. In it, he used the analogy of a person finding a watch to argue for the existence of God. If you found a stone lying on the side of the road, you may not think twice about it. But, if you found a watch, with all of its intricate machinery working in harmony, you would conclude that an intelligent being created it. Likewise, the order and complexity of the universe suggest that it, like the watch, was designed.

> If you found a stone lying on the side of the road, you may not think twice about it, but if you found a watch, with all of its intricate machinery working in harmony, you would conclude that it was created by an intelligent being.

Natural Theology was required reading for Darwin while he was at Cambridge, and at the time he trusted Paley's theories[5], only later to reject them in favor of his own theory of natural selection[6]. Paley's design argument had no rigorous standard for detecting design, other than being able to discern an object's purpose. Darwin's theory, based on the supposed evidence from artificial selection and Malthus' population essay, quickly took its place in the scientific community, and Darwin triumphantly concluded that "everything in nature is the result of fixed laws"[7]. However, Darwin failed to think critically about this statement, because, in order for laws to exist, there must be a *lawmaker*, so who might this Lawmaker be?

Intelligent design:

Darwinism suppressed design theory's flame until 1996, when Michael Behe stoked the fires with his book, *Darwin's Black Box*[8]. Current design theory is called Intelligent Design, to distinguish it from earlier design theories. ID theory involves the following two basic assumptions

1. Intelligent causes exist
2. These causes can be empirically detected

These assumptions lead to two questions, the first of which is "How do we know intelligent causes exist?" The answer lies in Romans 1:19 "what may be known of God is manifest in them, for God has shown it to them." Romans

1:19 says that it is impossible not to know about The intelligent cause. Only a fool would say in his heart "There is no God" (Psalm 14:1). The second question is "How are intelligent causes detected empirically?" In other words, how can you test for intelligent causes? Well, there are two ways; by observing specified complexity and irreducibly complex systems.

Specified Complexity:

Specified complexity is an unambiguous standard easily explained with an analogy: The probability of randomly pulling 7 letters from a bag of letters and spelling the word BIOLOGY is 1 in 26^7, or 1 in 8 billion. We would say the word BIOLOGY is specified (fits a recognizable pattern) and complex (1 in 8 billion chance of making the word using random processes). Anything has specified complexity if it has an extremely low probability of occurring by chance, and if it matches a discernible pattern.

Irreducibly complex systems:

This is a system composed of several well-matched, interacting parts contributing to the basic functioning of a system, wherein the removal of any one of the parts causes the system to cease functioning. One example of an irreducibly complex system is a mousetrap. It has several parts, but take away just one, and the mousetrap will not function properly. An almost infinite number of examples of irreducibly complex systems exist, both natural and manmade.

Strengths and Weaknesses of ID theory:

As we discuss the strengths and weaknesses of ID theory, we should keep in mind some ideas mentioned previously:

1. A good scientific theory accurately describes a large class of observations, making definite predictions about the results of future observations.
2. Any attempt to use the scientific method to prove what happened in the past is not real science but is speculation. All we can do scientifically is uncover evidences, which we manipulate to fit either the evolution or the creation framework.

ID theory is a good scientific theory:

The major strength of ID theory is that it is a good scientific theory. Take any gene in any genome for example. All genes show signs of specified complexity, because of their 3 base pair coding system. We can predict that if a mutation occurs in a gene, a change of function will result.

All organisms show signs of irreducible complexity. We can predict what loss of function will result from removing a certain component from any system, natural or manmade. If I remove the battery from my car, I can predict that the car will not start. If I clip a bird's wings, it will not be able to fly as well.

We know all organisms contain DNA, a molecule that is both specified and complex. Evolution cannot explain this information-rich material, and in fact "information-creating evolution is not empirical science at all because it has never been observed either in the wild or in the laboratory"[9]. Any discussion of how DNA formed is speculation, not real science. Fortunately, more and more scientists are recognizing that science simply cannot explain everything.

ID theory is not scientific creationism:

This may make ID theory sound "anti-Christian," but in reality it helps to separate science from speculation. To understand this better, let's look at a legal definition for scientific creationism[10]:

- ❑ The universe, energy and life were created from nothing
- ❑ Mutations and natural selection cannot bring about the development of all living things from a single organism
- ❑ "created kinds" of plants and organisms can vary only within fixed limits
- ❑ humans and apes have different ancestries
- ❑ Earth's geology can be explained by catastrophism, primarily a worldwide flood
- ❑ The earth is young—in the range of 10,000 years.

Scientists have mounds of evidence supporting each statement, but none can be proven scientifically, except possibly the third statement. To believe all of these statements as truth requires faith in God's word. Since faith is a gift, unbelievers would think it foolish to believe such things (I Corinthians 1:18).

Intelligent design, on the other hand, has a much more streamlined definition and involves two assumptions:

- ❑ Intelligent causes exist.
- ❑ These causes can be empirically detected (observed in other words) by looking for specified complexity

The first assumption is the major premise of intelligent design, supported by specified complexity, which is a good scientific theory. Now, it may seem that ID theory is "weak" because it has so few assumptions, but that is actually its strength. As a leader in the ID movement, William Dembski says "it doesn't speculate about a Creator or his intentions"[11]. ID theory simply says that intelligent causes exist and we can detect them.

Since Darwin first published Origin of Species in 1859, a creation/evolution battle has raged, with a large part of the battle waged by both parties using the same set of evidence to prove their way is right. Work done by creation scientists and creationist organizations such as ICR, Answers in Genesis, and many others has been important and necessary, because a Christian should "answer a fool according to his folly, lest he be wise in his own eyes" (Proverbs 26:5). Their work must continue until Darwinism has faded from the scene.

However, evidences used by both parties will only allow them to speculate about origins, since we weren't there to see what happened. Trusting either evolution or creation as truth is a matter of faith, not conclusive scientific evidence. The great thing about ID theory is that it does not speculate about origins, but it gets right to the point that a designer exists and we can detect evidence for him EVERYWHERE.

To *deny* that specified complexity proves intelligent causes exist, a person must deny the thoughts they think and the words they write were not created by an intelligent being, because even words and thoughts contain specified complexity. To *deny* that specified complexity proves intelligent causes exist, a person must either believe they are not an intelligent being or they don't exist, which is even more absurd than believing whales' ancestors were grazing land mammals.

> To *deny* that specified complexity proves intelligent causes exist, a person must either believe they are not an intelligent being or they don't exist.

ID expresses some of God's attributes:

Romans 1:20 says "*since the creation of the world, His invisible attributes are clearly seen, being understood by the things that are made.*" Specified complexity is concerned with information and information creating systems, and in living organisms, that means DNA, which contains the code words for making life. Therefore, specified complexity is analogous to God's word, because "*in the beginning was the Word, and the Word was with God, and the Word was God*" (John 1:1). God's word is definitely specified and complex!

Irreducibly complex systems show us how removing just 1 part can cause a system to function improperly. Irreducibly complex systems are analogous to the Body of Christ, for *"as the body is one and has many members, but all the members of that one body, being many, are one body, so also is Christ"*(I Corinthians 12:12).

Galileo once said *"the Bible tells us how to get to heaven, not how the heavens go."* Evidences about the earth's past do fit into the truths found in the Biblical account of origins, but the Bible is not a science textbook. What the Bible does do is show us that we are to *"consider His works"* and *"have dominion over them"* (Psalm 8), that His attributes are clearly seen through a study of His works (Romans 1:20), and that *"by faith we understand that the worlds were framed by the word of God, so that the things which are seen were not made of things which are visible"* (Hebrews 11:3).

Science and Christianity are inseparable, not because the Bible is a science textbook, but because science is a tool man uses to know God better, which is why so many of the best scientists and mathematicians were and are Christians.

ID should not be taught in public schools (but neither should origins):

During the 2003 biology textbook adoption process in Texas, there was a push to get intelligent design into the high school biology textbooks. A Zogby poll even indicated that 80% of Texans wanted intelligent design taught in public schools. However, instead of pushing for inclusion of ID, education reformers should instead push to get evolution and all discussion of origins out of the textbooks.

As we have discussed on multiple occasions, any discussion of origins of the earth, species, etc, is not real science, but is speculation, and to believe it as truth requires faith. A secular biology textbook should talk about matters related to science, not faith, and since public schools want so badly to be secular, then they need to get evolution out of their books.

Think of the huge waste of time involved in studying evolution. For example, Starr and Taggart's text contains 4 chapters on evolution and countless other evolutionary statements in other chapters.

> Everything in a biology course can be studied perfectly well without reference to origins.
> Randall Hedtke

According to Randall Hedtke, a retired public schoolteacher, "everything in a biology course can be studied perfectly well without reference to origins"[12], and I agree. If public schools want textbooks with evolution, then they should

include ID as well, especially since evolution cannot explain information creation, but ID can. However, I think they should have neither evolution nor ID,
because to believe ID, you have to believe that intelligent causes exist, which is
a faith-based belief.

ID Doesn't say who the designer is:

Darwin proclaimed "everything in nature is the result of fixed laws"[13], but
he obviously didn't realize laws don't just "appear," they are *created*. ID theory
proclaims there is a lawmaker, but my main concern with the ID movement is
that it does not specify who the lawmaker is.

One reason the lawmaker is not specified is because intelligent design can
describe both God-made and man-made causes. Every word spoken by a
human shows signs of specified complexity, and man builds all kinds irreducibly complex things. ID theory can apply to any organized matter, God-
made or otherwise.

Another reason intelligent design does not specify the designer is because in
the secular world, if you say "God is the designer," you will likely end up with a
"religious battle" on your hands. This makes it difficult to discuss the theory's
scientific aspects. However, remember that ID's major premise is that "intelligent causes exist," which is a pretty vague description, but anyone who is
thinking critically will ask questions like "Do these intelligent causes have a
name?", and we need to have an answer for them.

Darwin used to believe in design theory, and in the Designer. However, he
rejected the Designer in favor of his own theory. Darwin deceived himself
because we know who the real Designer is, because what may be known of
Him is manifest in us (Romans 1:19). According to the Bible, it doesn't take
faith to know that God exists. Everyone knows who He is. It just takes faith to
trust Him with your life, and that faith only comes from Him.

Christians should use ID theory as a way to explain information creation,
which is going to be the downfall of Darwinism, since it cannot explain information creation. We should also boldly proclaim who the Designer is. Just like
any tools made by man though, we can use them to bring glory and honor to
His name, or we can use them to generate confusion and destruction.

ID theory should not be taught in public schools, because they won't proclaim who the real Designer of living organisms is. Even if they do, they will
likely exchange the truth and say the intelligent cause was a "seed" from
another planet, space aliens, alla, Zeus, Captain Crunch or the Michelin Man.

Summary

We should teach the message about intelligent design and the fallacy of evolution through Christian churches, Christian schools, and Christian families, and when the world sees the Truth, they will want to be free as well. Until public schools stop promoting that the fear of the Lord is the beginning of superstition[14] and start promoting the Truth that the fear of the Lord is the beginning of wisdom (Proverbs 1:7), they will not give students an excellent and proper science education.

> An excellent science education sticks to the facts, and discusses science history and the significance of men who studied God's word and works together.

An excellent science education sticks to the facts, and discusses science history and the people who studied God's word and works together. Students learn that it was because of scientists who trusted The Intelligent Designer, not despite them, that science has flourished. God's Word directed their study of His works, and that is why they were so good at what they did.

PART IV
What Happens When You Believe a Lie

10 PICK ONE
Evolution and Scripture Cannot Both Be True

During this course, we have studied Darwin's works directly. We found that his major premise in writing *Origin of Species* was to "show that the species had not been separately created, and secondly, that natural selection had been the chief agent of change"[1]. To explain natural selection, Darwin didn't look to anything natural, but instead combined ideas from artificial selection (breeding) and the faked data of Thomas Malthus. Until his time, most scientists believed the species were separately created as described in Genesis 1, but Darwin cast those views aside.

In retrospect, if Darwin based his theory on facts and direct observation, then it would be much more credible, but he based it on lives, which makes it *incredible* that it has lasted so long. In an attempt to disprove the Genesis account, Darwin believed lies.

What happens when someone believes lies? Well, all of Scripture is God-breathed (2 Timothy 3:16), and the entirety of His word is truth (Psalm 119:160). Therefore, if someone exchanges the truth of God's word for a lie in one part of Scripture, this will have a cascading effect, exposing the lie in many other parts of Scripture. One lie is like a tiny snowball at the top of a tall mountain. It seems harmless enough at first, but then the first lie collects another, and another, and soon an unstoppable avalanche has formed, leaving incredible destruction in its path.

Genesis 1 talks about how God created the animals "according to their kind" or separately. Since Darwin's major premise for *Origin of Species* contradicts Genesis 1, we will find many other places where it contradicts Scripture. Two significant places include Scriptures referencing "the wages of sin" and "the meek."

Evolution vs. the Wages of Sin

God created man in His image (Genesis 1:26). He named the first man Adam, and the first woman, Eve. It didn't take them long, however to disobey God. God made them a beautiful garden to work in, full of all kinds of trees to gather food from. However, he told them not to eat from the tree of knowledge of good and evil, because if they did, they would die (Genesis 2:17). So, what do they do? Well, they go and eat from the one tree they weren't supposed to eat from. Of course, it didn't help that Satan was putting doubts into Eve's mind by asking questions such as "Did God really say you're not supposed to eat from that tree?" (Genesis 3:1).

Ultimately, it was man who decided to eat from the tree, and his disobedience is what *sin* is about. Sin means to "miss the mark." Adam and Eve missed the mark when they disobeyed God's commands. Scripture teaches the penalty or "wages" of sin for all mankind is death (Romans 6:23). What we "earn" for our disobedience is death. However, the gift of God is eternal life in Christ Jesus our Lord (Romans 6:23). Jesus paid the penalty for us by dying for our sins (Matthew 26:28, Hebrews 9:27-28).

> Since Scripture teaches that *sin brought death* (Romans 6:23), this is in direct contradiction to evolution, which teaches that death brought man.

Now, what does this have to do with evolution? Well, since Scripture teaches that *sin brought death* (Romans 6:23), this is in direct contradiction to evolution, which teaches that death brought man. In his concluding paragraph in *Origin of Species*, Darwin stated "Thus, from the war of nature, from famine and *death*, the most exalted object which we are capable of conceiving, namely the production of the higher animals, directly follows."[2] God's word teaches that death is a penalty, while evolution teaches that death is a creator.

Scripture reveals that the billions times billions of deaths required by evolution to form humans are an impossibility. Romans 5:18-19 says

Therefore, as through one man's (Adam) offense judgment came to all men, resulting in condemnation, even so through one Man's (Jesus) righteous act the free gift came to all men, resulting in justification of life. For as by one man's disobedience many were made sinners, so also by one Man's obedience many will be made righteous.

Not only does evolution attempt to discredit the Genesis account, it also attempts to discredit Christianity itself, which Darwin considered a "damnable doctrine"[3]. If death brought man, which is what evolution proposes, then there is no place for sin, and there is no reason for Jesus to die for our sins. It would be pointless for Him to die for our sins if death was not the penalty for sin. Belief in evolution is more than just disbelief in special creation by God, it is disbelief in Christianity itself. An excellent resource on this topic is a video titled "*Evolution and the Wages of Sin*" by the Institute for Creation Research (www.icr.org).

Evolution vs. the meek

Evolution theory speaks of a "struggle for existence" and a "survival of the fittest." According to evolution theory, the strongest, most superior genetic stock will always conquer. In the *Descent of Man*, Darwin applied his theory to humans, and clearly distinguished the "inferior" as the poor classes, when he said

All ought to refrain from marriage who cannot avoid abject poverty for their children; for poverty is not only a great evil, but tends to its own increase by leading to recklessness in marriage. On the other hand, as Mr. Galton has remarked, if the prudent avoid marriage, whilst the reckless marry, the inferior members tend to supplant the better members of society.[4]

Note also that the subtitle to Origin of Species is "*or the Preservation of Favored Races in the Struggle for Life*," which most modern printings of the book leave off. Clearly, Darwin believed the "favored races" were the wealthy.

> If evolution is true, the Bible should support it with verses expressing how the healthiest, wealthiest, most powerful people will dominate the earth.

If evolution is true, the Bible should support it with verses expressing how the healthiest, wealthiest, most powerful people will dominate the earth. Of course, the Bible contradicts this idea. Let's look at a few Scriptures:

Psalm 8:2 "Out of the mouths of babes and nursing infants you have ordained strength, Because of Your enemies, that You may silence the enemy and the avenger"

Matthew 5:5 "Blessed are the meek, for they shall inherit the earth."

Philippians 2:3 "Let nothing be done through selfish ambition or conceit, but in lowliness of mind let each esteem others better than himself"

Isaiah 40:30 "Even the youths shall faint and be weary, and the young men shall utterly fall, but those who wait on the Lord shall renew their strength."

Matthew 11:29 "Take my yoke upon you and learn from Me, for I am gentle and lowly in heart, and you will find rest for your souls. For My yoke is easy and My burden is light."

Galatians 3:28 "There is neither Jew nor Greek, there is neither slave nor free, there is neither male or female; for you are all one in Christ Jesus"

I Samuel 17:45 "So David prevailed over the Philistine with a sling and a stone, and struck the Philistine and killed him."

Jesus does not "categorize" people like we do. To Him, you are either following Him or not, and it doesn't matter if you are a man or a woman, good athlete or bad, smart or dumb, rich or poor, minority or majority, inferior or superior. God desires righteousness, humility and obedience to His will. He doesn't need us wasting time making up theories that attempt to justify one group as superior to another. He wants us to consider others better than ourselves, not consider ourselves better than others, which is exactly what happens when someone applies evolutionary theory to society.

The Bible says "all have sinned and fallen short of the glory of God" (Romans 3:23), and that "none are righteous" (Romans 3:10), and in our unrighteousness we have suppressed the truth (Romans 1:18). As sinners we have to humble ourselves and consider others better than ourselves. Now, does that mean if we run a mile race with another person and we beat him fair and square that we should say to him "you are really better than I am"? Of course not. What it does mean though is if your neighbor is starving and you have food to spare, you should share it, or if you want to go fishing and your wife wants a break, you should give her a break. God does not want us to set up our own little kingdoms and think we are "somebody." He wants to display His power and His love through us, and he cannot do that if we are too busy trying to be better than everyone else.

As a little shepherd boy, David understood the meaning of humble obedience, as he said to the mighty Goliath "You come to me with a sword, with a spear, and with a javelin. But I come to you in the name of the Lord of hosts, the God of the armies of Israel, whom you have defied."

Chapter 10 Summary:

Either the Bible is wrong, evolution is wrong, or they are both wrong. Both cannot be right. It is impossible that death is both the creator of man (through evolution) and the penalty for man's sin. If evolution is true, then Christ's death is meaningless. The two ideas completely contradict each other.

> The theory of natural selection is really more of an excuse for a bad attitude towards God and man than it is a theory of any scientific use. To many, it is an excuse for sin.

Evolution does not encourage a humble heart, but rather encourages an unrighteous heart and an "I'm better than you" attitude. In a nutshell, evolution promotes sin. God wants us to be meek, or humble, and to have the attitude of a servant. Jesus himself was descended from the line of David, the humble shepherd boy who trusted in the Lord for victory, not in himself.

The United States of America was founded by Christians seeking to worship the Lord freely and without oppression from a king who thought he was better than everyone else. I believe that God has blessed America because so many of those who founded the country were seeking righteousness. It is because of Him, not in spite of Him, that we are the healthiest and wealthiest nation the world has ever seen. However, our public schools and most universities do not teach about America's Christian Heritage anymore, which leads people to believe America is great because of man, not because of God. This is a huge error, because it is an exchange of "the glory of the incorruptible God into an image made like corruptible man, and birds, and four-footed animals and creeping things." It is the attitude that "God isn't God, I am god!", and the theory of natural selection has undoubtedly steered many down the path of deception into thinking they are a god. I think it is fair to conclude that *the the-*

ory of natural selection is more of an excuse for a bad attitude towards God and man than it is a theory of any scientific use. To many, it is an excuse for sin..

11 MASKING SIN
The Effects of an Evolutionary Worldview on Society

Darwin wrote *Origin of Species* to disprove the Genesis account[1], and he talked honestly and openly about his anti-Christian and anti-God beliefs. I believe in right and wrong, good and evil, and I believe there is one God, and I believe there is one devil. The devil, or Satan, first appeared in Genesis Chapter 3, tempting Eve to sin by eating from the one tree God commanded her not to eat from. Now, did God only punish Satan for tempting Adam and Eve to sin, letting them off the hook because Satan initiated the tempting? No! Adam and Eve received the penalty of death for their sin.

My point here is that while Adam and Eve believed Satan's lie, they *chose* to sin. Surely, it didn't help that Satan was there putting questions in their minds, but ultimately, it was Adam and Eve's own fault that they sinned.

Compare the story of Adam and Eve and Satan to Darwin and what many people call *Darwin's henchmen*. Surely, Darwin's theory has tempted many people to commit some horrific acts, but we must keep in mind that it is people who sin, not theories. We are all sinners, whether we are followers of

Darwin or followers of Christ, and many sins have been committed in the name of natural selection, just as many sins have been committed supposedly in the name of Christ.

God allows Satan to test our faith (Genesis 3:1-4, Job 1:6-12, Matthew 4:1-11), but ultimately we are to blame for our sins. With this in mind, let's see how Satan has used the theory of natural selection to tempt many to suppress God's truth and exchange it for the lie. I think it is because of, and not in spite of the evolutionary worldview and it's "I'm better than you" mentality that many refer to the 20th century as the bloodiest century in the history of the world.

> God allows Satan to test our faith, but ultimately we are to blame for our sins.

To study the effects of an evolutionary worldview on society, we will look at the ideas of not just Darwin, but also Thomas Malthus and Francis Galton. We will discover how the evolutionary worldview influenced the Belgian colonists of Africa, Hitler and the Nazis, Margaret Sanger and Planned Parenthood, and America's public schools.

Darwin, Malthus and Galton.

Thomas Malthus (1766-1834)

As discussed in prior chapters, Darwin used Malthus' makeshift population studies to support his theory. Malthus was also a proponent of the "I'm better than you" attitude, and his book, *An Essay on the Principles of Population*, published in 1798, promoted repressive legislation which worsened conditions for the poor in England. "*Malthusianism*" achieved its greatest triumph in 1834 with a new law providing for the institution of workhouses for the poor. Sexes were strictly separated to curb the otherwise inevitable over-breeding and subsequent take-over by the lower classes of the self-appointed "better people" [2].

Charles Darwin (1809-1882).

Darwin's book, *On the Origin of Species* was subtitled *Or the Preservation of Favored Races in the Struggle for Life*. His book, *The Descent of Man*, added to the idea of "favored races," as he concluded "there can hardly be a doubt that we are descended from barbarians" [3]. Scripture gives the

> Our sin nature, if followed instead of God, is what makes us descend *to* barbarians, rather than descend *from* barbarians.

opposite picture, in that our sin nature, if followed instead of God, is what makes us descend *to* barbarians, rather than descend *from* barbarians.

God's word teaches us "*The fear of the Lord is the beginning of knowledge*" (Prov. 1:7), and to "*Take firm hold of instruction, do not let go; keep her, for she is your life*" (Prov. 4:13). If we follow our sin nature instead of Christ, not retaining God in our knowledge, then He will give us over to a debased mind (Romans 1:28). The *Lord of the Rings Trilogy* by J.R.R. Tolkien provides an excellent example in the character of *Smeagol*, or *Gollum*, on the effect that following our sin nature can have in making us descend *to* the state of a barbarian.

Francis Galton (1822-1911).

Galton was an English psychologist and a half-cousin of Darwin. Fascinated by Darwin's theory, he sought to extend it into a concept of deliberate social intervention, which he held to be the logical application of evolution to the human race. Galton was by no means satisfied to let evolution take its course freely. Having decided to improve the human race through selective breeding, brought about through social intervention, he developed what he called "Eugenics." The principle of eugenics was that by encouraging better human stock to breed and discouraging reproduction by less desirable stock, the whole race could be improved[4].

While the idea of "good genes," or eugenics, is not entirely bad, it can be applied towards destructive ends. As we have said before, theories do not sin, people do. A good application of eugenics would be if a husband and wife knew their genetic makeup would result in a child with a lethal disorder, they could use that information to decide whether they should have their own children or adopt. It should be a way for Christians to carry out God's mandate described in Proverbs 14:8 "The wisdom of the prudent is to understand his way."

Eugenics, however, should never, ever be a tool governments use to decide who can and cannot have children, or even worse, who can and cannot live. Unfortunately, as we will see next, this is exactly what some have used the ideas of

> Theories do not sin, people do.

Malthus, Darwin, and Galton for. And as we study the next four examples, I encourage you, once more, to keep in mind that it is people, not theories who sin.

Genocide in Africa.

> *'Hey journalist! I will kill you, journalist!' The boy looks at me*
> *with blood-colored eyes-he could not be older than nine or ten-*
> *and waves the muzzle of his AK-47 in my face. He is drugged-up,*
> *gone. He wears a red t-shirt and a sardonic grin, his camouflage*
> *pants new and stiff-looking, hanging from his body like oversize*
> *pajamas. The back of the truck is full of others like him, child sol-*
> *diers: a 13-year old with a rocket launcher, a couple more*
> *teenagers with assault rifles. The driver, another kid, grinds the*
> *gears of a Toyota, trying to learn hands-on how a transmission*
> *works. I pause on the side of the street, intent on not showing*
> *fear. Fear is what they want, these children in control of Bunia[5].*

This is an excerpt from a story by a brave journalist who was in Bunia, in the
Democratic Republic of the Congo, on her way to do a story on the vanishing
population of mountain gorillas in the area. It seems however, that this part of
Africa, which includes Rwanda, Uganda, and Burundi, has not always been
such a place of violence, and in particular, genocide. However, since about
1960, around 5,000,000 people have been killed because of race-related issues.

So what caused the changes? Again, while people are ultimately responsible
for their sins, the lie of Darwinism undoubtedly played a role in the changes.
The journalist describes the colonization of DRC like this:

> *The Hema and Lendu peoples lived relatively peacefully together*
> *for centuries in Ituri until the Belgians colonized the Congo in*
> *1908 and declared the Hema as **racially superior** [emphasis*
> *mine], thus relegating the Lendu to a permanent lower-class sta-*
> *tus.... Enmity grew between the Hema and Lendu, but tensions*
> *escalated to wholesale violence after Western powers discovered*
> *that Ituri is home to some of the largest gold reserves in the*
> *world.[6].*

The plight of the Hutu and Tutsi peoples, who mainly live in Uganda and
Rwanda, is strikingly similar. According to Gregory Mthembu-Salter:

> *Germany ruled Rwanda until the end of WWI, when the colony*
> *was transferred to Belgium. Both the Belgian colonial adminis-*
> *tration and the new Roman Catholic missionaries, called the*
> *White Fathers, who were the most active proselytizing body in*

the territory, were race-obsessed. They treated the highly complex issue of the difference between Hutus and Tutsis, which involves factors of heredity, class, and social obligation, as straightforwardly ethnic. The White Fathers and the administration swiftly concluded, on flimsy evidence, that Tutsis and Hutus were of completely separate ethnic origin and the Tutsis were the Hutus' natural masters. They believed the Tutsis were "sub-Aryan" with ancient Christian ancestors, while Hutus were merely ordinary Bantus, worthy to be nothing more than 'hewers of wood and drawers of water'[7].

In his internet book, *The Ethnic Trap!*, native Rwandan Benjamin Sehene said:

The shape of one's nose, his height, his father's name, the place of his birth, are proof of his identity, his worth, the distinction and separation of oneself, the proof of one's position and function. One is never Rwandese, he is either Tutsi, Hutu or Twa. Although Rwanda is one of Africa's oldest nation states, ethnicity, not nationality, is what defines the individual. Rwanda has indeed fallen into the ethnic trap; first set by nineteenth century European ethnologists and missionaries still in the throes of Darwinism, before being adopted and sustained by Hutu ideologists when they took power in what would be termed the 'social revolution' in 1959, followed by thirty-five years of ethnic tyranny culminating in the 1994 genocide of Tutsis.[8]

From April to July of 1994, Hutus massacred over 800,000 Tutsis and moderate Hutus. In June, "bloated, dead bodies floated down the Nyabarongo and Akagera river towards Lake Victoria, at the rate of two bodies per minute"[9]. Darwinism, especially the results of its application, is evident in the genocide of millions of Africans, which continues today.

> Rwanda has indeed fallen into the ethnic trap; first set by nineteenth century European ethnologists and missionaries still in the throes of Darwinism.
> *Benjamin Sehene*

Hitler and the Nazis

While most people are unfamiliar with the Darwinism-influenced atrocities occurring in Africa, Hitler's philosophies are more well-known. Hitler's "view of life was strongly Social Darwinist, society being seen as an arena in which individuals and groups were engaged in a ceaseless struggle to assert their superiority by force and cunning"[10]. Hitler did not think it was fit for the "superior" to mingle with the "inferior."

"Inferior" to Hitler usually meant Jewish, but it also meant anyone who disagreed with him. Hitler thought he would "help God" get rid of sin by not permitting superior and inferior to have children together. For example, in his book, *Mein Kampf*, or *My Struggle*, Hitler said such "interbreeding" was "a sin against the will of the Eternal Creator"[11]. His Darwinist views also led him to say "that it is the strongest who must triumph and that they have the right to endure"[12]. I guess Hitler never read the story of David and Goliath, and I guess he wasn't around

> Hitler thought he would "help God" get rid of sin by not permitting superior and inferior to have children together.

in 1980 when the U.S. Hockey team beat an obviously "better" Russian team. Hitler's eugenic, sterilization, and racial hygiene programs were responsible for the deaths of over 11 million men, women, and children.

Margaret Sanger and Planned Parenthood

"Human weeds, reckless breeders, spawning human beings who never should have been born"[13] are a few of the ways Margaret Sanger, founder of Planned Parenthood, described those she considered "inferior" humans. Sadly, unbelievably, she admitted that her life's purpose was to promote birth control[14], resulting in the murder of almost 45 million babies in America[15] and up to 2.5 billion worldwide[16].

Few know of her attitude or her priorities, and according to George Grant, author of *Killer Angel*, "her faithful minions have managed to manufacture an independent reputation for the perpetuation of her memory"[17]. Grant describes her as not only the "progenitor of the grisly abortion industry," but the "patron of the devastating sexual revolution"[18]. It is ironic that Sanger, who considered the poor as "inferior," had an "impoverished and isolated" upbringing, in a home with an atheistic, hateful father.

However, it was not poverty, but hate, that shaped and molded her the most. She studied intently the works of Thomas Malthus, the same man whose ideas formed the basis of Darwin's theory of natural selection. She liked his ideas of population control and his desire to refrain from Christian charity, which, to him, simply aggravated the "problems" related to the poor. She grew tired of her marriage to William Sanger, and fell in love with Havelock Ellis, a disciple of Francis Galton and the Eugenics movement[19].

In the 1920's, her magazine, *Birth Control Review*, predecessor to the *Planned Parenthood Review*, regularly and openly published the racist articles of Malthusian Eugenicists. In April of 1932, the magazine outlined Margaret's "Plan for peace," which called for coercive sterilization, mandatory segregation, and rehabilitative concentration camps for all "dysgenic stocks"[20]. In April of 1933, the Review published a shocking article entitled *Eugenic Sterilization: an Urgent Need*. It was written by Margaret's close friend and advisor, Ernst Rudin, who served as Hitler's director of genetic sterilization and had a prominent role in the establishment of the Nazi Society for Racial Hygiene[21].

> The bottom line is that Margaret self-consciously organized the Birth Control League—and its progeny, Planned Parenthood—in part, to promote and enforce the scientifically elitist notions of White Supremacy
>
> *George Grant*

According to Grant "The bottom line is that Margaret self-consciously organized the Birth Control League—and its progeny, Planned Parenthood—in part, to promote and enforce the scientifically elitist notions of White Supremacy. Like the Ku Klux Klan, the Nazi Party, and the Mensheviks, Margaret's enterprise was from its inception implicitly and explicitly racist"[22]. Gee, thanks Margaret's dad, for teaching your daughter how to hate.

America's public schools:

With the help of people like Margaret Sanger, the early 1920's found social patterns in America in a state of chaos. Who would dominate the American culture—the modernists or traditionalists? Journalists were looking for a showdown, and they found one in the Dayton, Tennessee courtroom in the summer of 1925[23]. There, a jury was to decide the fate of John Scopes, a high school biology teacher charged with illegally teaching about Darwinism. Instead of focusing on whether Scopes had broken the law, the two lawyers-Clarence Darrow was the defense attorney, William Jennings Bryan was the prosecuting attorney-focused on the creation/evolution controversy, and whether laws forbidding the teaching of evolution were constitutional. While John Scopes was eventually fined $100 (judges later reversed this decision), the press described the trial as a victory for Darwinism.

Inherit the Wind, a movie based on the trial, hit the screens in 1960, just over 100 years after publication of the first edition of *Origin of Species*. Philip Johnson describes the real trial and movie as

> *a publicity stunt by the ACLU, but Broadway and Hollywood converted it into a morality play about religious persecution in which the crafty defense lawyer Clarence Darrow made a monkey of the creationist politician William Jennings Bryan, and in the process taught the moviegoing public to see Christian ministers as ignorant oppressors and Darwinist science teachers as heroic truth seekers. As the 20th century came to an end, science and history teachers were still showing Inherit the Wind to their classes as if it were a fair portrayal of what happened in Dayton, Tenn in 1925.*[24]

As a result of the Scopes Trial and *Inherit the Wind*, in the early 1960's evolution theory received much added emphasis in biology textbooks[25]. These textbooks usually describe evolution as their "unifying theme." For example, the biology textbook by Starr and Taggart that we have referred to several times is titled *Biology: The Unity and Diversity of Life*, where they describe evolution as the "unifying theme."

Once again though, this time in America's classrooms, the "I'm better than you" attitude of Darwinism had murderous results. They are best described in a letter to the Texas State Board of Education by Darrell Scott, father of Columbine High School victim Rachel Scott:

> *You should understand that some students, surely not many, but some, will take the concept that we are 'nothing but animals' and 'survival of the fittest' as justifications for heinous behavior. Hitler did it. Stalin did it. And Eric Harris and Dylan Klebold, killers of my daughter and twelve other students and teachers at Columbine High School did just that. Harris wore a 'Natural Selection' t-shirt on the day of the killings and both had taken the evolutionary mindset to the conclusion that they needed to help the process along. They made remarks on video about helping out the process of natural selection by eliminating the weak. They also professed that they had 'evolved' to a higher level than their classmates.*[26]

Chapter 11 Summary

If you believe Darwinism is a good theory, applicable to man and animal, then consider a few questions. Which of your friends or family members are you "better" than? Who is the most "evolved" group, Whites, Blacks, Browns, Reds, or Yellows? Your Darwinist beliefs require you to choose an answer to such questions.

Darwin's theory is based on lies, and is not a real scientific theory, but is more of an excuse for a bad attitude or opinion that says "I'm better than you, now get out of my way!" Here are a few other conclusions:

- We should quickly eliminate any attempts at creating social programs promoting an "I'm better than you" attitude. Programs containing the words eugenics, euthanasia, sterilization, etc. should be red flags for us, and we should stop them in their tracks, or we may end up with another holocaust.

- One man can make a difference, because in all the situations I presented relating an evolutionary worldview and murder of innocent lives, just one or two people were the ringleaders. If you don't think it matters whether schools teach evolution, then you are deceived, because the more exposure students receive, the more chance there is of one person being tempted by its lies. Schools either need to eliminate evolution from the textbooks that they use, or if they do discuss it, discuss that it is merely speculation and not a real theory.

- Evolution theory does not sin, people do. Evolution theory simply gives some people an excuse to have an "I'm better than you" attitude. We have looked at verses like Luke 10:27 and Philippians 2:3-4 to understand that God's word totally contradicts the evolutionary mindset. James 2:8-15 is another excellent passage about what our attitude towards others should be like.

> *If you really fulfill the royal law according to the Scripture, 'You shall love your neighbor as yourself', you do well;* **but if you show partiality, you commit sin** *[emphasis mine], and are convicted by the law as transgressors. For whoever shall keep the whole law, and yet stumble in one point, he is guilty of all. For He who said, 'Do not commit adultery,' also said, 'Do not murder'. Now if you do not commit adultery, but you do murder, you have become a transgressor of the law. So speak and so do as those who will be*

judged by the law of liberty. For judgment is without mercy to the one who has shown no mercy. Mercy triumphs over judgment.

Malthus, a clergyman, and Hitler, who thought he was helping God, must have skipped over this part of the Bible. As you continue life's journey, constantly check your attitude towards others, and if your heart is not "How can I serve you?" but rather "How can you serve me?", then get ready to be judged without mercy.

> Malthus, a clergyman, and Hitler, who thought he was helping God, must have skipped over James 2:8-15.

12 RENEW YOUR MIND
Education

We have come to the last, but definitely not the least important, chapter in our story. Thankfully, our story has a beautiful ending, like a spectacular sunset at the end of a busy day.

Lakeshore Baptist Church (Fig. 12.1) in Lakeshore, Mississippi, began in 1911 in the home of R. C. Crysell[1]. However, on the morning of August 29th, 2005, Hurricane Katrina slammed into the tiny community, reducing beautiful little Lakeshore Baptist Church to a slab. Many felt like giving up, but many more believed they could and should rebuild.

What followed was one of the most amazing acts of restoration America has ever witnessed, as volunteers poured in from across the country to restore not only Lakeshore, but hundreds of towns and communities up and down the Gulf Coast. Many of these towns had little government help and little money of their own, but volunteers brought needed supplies and helped "pay the debt."

Figure 12.1. Lakeshore Baptist Church, an image of Creation, Fall, and Redemption. Top: Lakeshore Baptist before Hurricane Katrina. Middle: On August 29, 2005, the church was reduced to a slab. Bottom: Lakeshore Baptist, 1 year after Katrina, is a fully functioning church and a hub of activity in the community.

What does this story have to do with education? Well, it has everything to do with education, especially the Creation/Fall/Redemption message contained within. God created, man fell, Christ paid the debt for our sin, and whether we like it or not, we are currently participating in the restoration of His kingdom on earth. The restoring of man's mind through a Christian education is a significant part of that work.

> We are currently participating in the restoration of His kingdom on earth, and the restoring of man's mind through a Christian education is a significant part of that work.

Creation, Fall and Redemption are not only fundamental turning points of biblical history, they are marvelous diagnostic tools.[2] Let's turn our attention now to the history of education and watch how the Creation/Fall/Redemption story has been told repeatedly over the last several hundred years, beginning with the creation of Classical Christian Education as a tool for restoring the mind of man. We will conclude with a message of hope for the future, not just for education, but for mankind.

Creation of Classical Christian Education

Our story begins late in the 12th Century, when a phenomenon unique to Europe appeared, *the university*. Originally, these schools owned no real estate, but were instead an association of teachers or students. Although not always theologically or scholarly accurate, what under girded the university was the unification of all subjects by an all-encompassing worldview. Christianity provided the bridge connecting all departments of study[3].

Classical Christian Education produced some of the greatest scientific and mathematical minds the world has ever known, men like Galileo, Kepler, Mercator, Newton, Leibniz, the Bernoulli's, Euler and Maury, to name a few. To these men, Christianity and science were thought to be completely compatible and mutually supporting[4], as evidenced in their writings. For example, in the summer of 1760, when describing vision to a German princess, Leonard Euler exclaimed:

> *I am now enabled to explain the phenomena of vision, which is, undoubtedly, one of the greatest operations of nature that the human mind can contemplate. Though we are very far short of a perfect knowledge of the subject, the little we do know of it is more than sufficient to convince us of the power and wisdom of the Creator. We discover in the structure of the eye, perfections which the most exalted genius could never have imagined.*[5]

Notice how he writes confidently about God, bringing glory and honor to His name, acknowledging Him as the creator of vision.

The Fall of Education

In any Creation/Fall/Redemption story, mankind inevitably falls after an attempt to replace God's word with man's ideas, and the story of the fall of education is no different. Classical Christian Education dominated Western Culture for hundreds of years, but a slight "stumble" occurred in the 17th century, with the overemphasis of Aristotle's ideas. Galileo noticed it in the Catholic Church, and Francis Bacon noticed it in England's universities. Bacon referred to the overemphasis of Aristotle as a "disease of learning," believing Paul's message in I Tim. 6:20, warning against false knowledge, was prophetic for his day as well[6]. As we learned in Chapters 1 and 2, Bacon sought to advance learning by studying God's word and His works together, restoring the Christian attribute of unity and diversity in education.

Educators liked Bacon's idea of advancing learning, but somewhere along the way, they forgot the part about studying God's word alongside His works. In America, many educators replaced an overemphasis on deductive reasoning in the form of Aristotle with an overemphasis on inductive reasoning in the form of "experience." And Darwin's *Origin of Species*, which British educational analyst Melanie Phillips describes as having a "shattering impact"[7], shoved education over a cliff.

In America, John Dewey kept education falling fast, although he did have some good thoughts, such as this excerpt from his 1916 book *Democracy and Education*:

> On one hand, there is the contrast between the immaturity of the new-born members of the group—its future sole representatives—and the maturity of the adult members who possess the knowledge and customs of the group. On the other hand, there is the necessity that these immature members be not merely physically preserved in adequate numbers, but that they be initiated into the interests, purposes, information, skill, and practices of the mature members: otherwise the group will cease its characteristic life.[8]

This seems reasonable enough, assuming that by "the group," Dewey meant the children's immediate family. But that's not what he meant. According to Melanie Phillips, his educational philosophy was

> *Explicitly linked to an egalitarian (meaning equal results), socialist agenda, its intellectual roots owing much to a way of thinking which viewed human beings as ahistoric members of an animal species, adapting themselves to their environment and their environment to themselves. Under this egalitarian banner, what he institutionalized in education was at root a savage doctrine of individualism which would effectively abandon children to a world without culture.[9]*

Dewey based his doctrines on the idea of personal growth, where values were re-invented by each child. Parents and family, religion and history, were removed from the equation, supposedly to give children a "fresh canvas" to paint their own picture. Phillips goes on to say

> *Dewey could not tolerate any distinctions between individuals; in his view, any division between the educated and uneducated would lead to unacceptable divisions between them. All classifications and stratifications were to be ruled out, as were any inner or spiritual qualities which couldn't be shared. Not surprisingly, this train of thought led directly to the destruction of learning and of culture, not to mention the idea of moral authority itself, the understanding of common and eternal values that needed to be transmitted through that culture..[10]*

Dewey's system involved unity and diversity, but it was unity in the form of egalitarianism, and diversity in the form of individualism. It was all about man and his relationship with himself, rather than about man and his relationship with God.

> John Dewey's philosophy was all about man and his relationship with himself, rather than about man and his relationship with God.

Dewey thought that if he could make everyone "equally fit," then, according to Darwinism, everyone would survive! Amazingly, Dewey's weird ideas still abound in America's public schools and in American society. Thankfully, they won't last forever, because Christ will eventually restore His kingdom to perfection.

Restoration

It was a really bad idea to use hydrogen gas to make the Hindenburg lighter than air, because hydrogen is very flammable. It just took one spark, and seconds later on May 6, 1937, the Hindenburg was gone (Fig. 12.2). Likewise, it was a really bad idea for Dewey to apply Darwin's ideas to education. America's public schools are currently filled with hydrogen gas in the form of anti-God, humanistic and individualistic ideas. The spark that will ignite secular education in a huge fireball is the Classical Christian method.

Figure 12.2. The fate of godless secular education will be similar to the Hindenburg's.

In America, Christians started giving up on public education in increasingly significant numbers around 1980[11]. The number of home schools and private Christian schools has been on the increase since then, and is not showing signs of slowing down.

Like the discovery of a long-sunken treasure, Classical Christian education has been recovered. At first, there was a lot of muck to scrape off, but books like Cornelius Van Til's *Essays on Christian Education* and Douglas Wilson's *Recovering the Lost Tools of Learning* proved to be powerful cleaners. Classical Christian Education is now shining brighter than it has in hundreds of years, but there is still a ways to go before it is restored to its original beauty and luster.

I first learned of Van Til while reading *Mathematics, Is God Silent?* by James Nickel, and while *Essays* is a difficult read at times, it is well worth the effort. In it Van Til describes knowledge as a whole collection of facts, subject to interpretation. Who would be in charge of interpreting the facts, the Christian or the non-Christian? Van Til said "A basic assumption of anti-theistic thought is that man interprets reality for himself"[12], which is exactly what Dewey promoted.

While creativity of human thought is a foundational principle of nontheistic education, Van Til explained that creativity of God is the foundational principle of theistic education[13]. In order to interpret reality, or anything for that matter, one must find in it the plan of God. Once the plan is found, truth is found[14]. The non-Christian view would be that human thought is independent of God and man-centered, rather than God-centered.

> Creativity of God is the foundational principle of Christian education.
> *Cornelius Van Til*

A main feature of Christian education is the connection between the whole of History and God's absolute authority over His creation[15]. Since God is the author of History its relation to God is the most important thing to know about Him[16]. The Christian man should be at the center of the curriculum, serving as a reminder that God has made the environment subject to man, and not man subject to the environment[17] as Darwinists would want us to believe. God designed us to take dominion (Genesis 1:28), and gave us the ability and the responsibility to rule over His creation. God did not design His creation to rule over us, and we are not in a constant battle to survive it, and no matter how hard they try, fire ants, rats and pigeons will never rule the world!

A proper Christian education will teach our children how to be good rulers over Creation. Every course, from math to music, has a biblical, Christian message to tell. We must constantly guard against the desire to teach religion indirectly, saying to ourselves, "this lesson has no Christian application." It takes more thought to find the Christian message within every lesson of every course, but if we take the easy way and save religion for Sunday mornings, then everywhere and always will we spend less time teaching religion directly[18].

In June of 1999, the Southern Baptist Convention adopted a resolution promoting Christian education, urging churches to "enter into and vigorously support education programs for children and youth that will provide biblically-based education"[19]. They continue pushing hard, and in 2006 put forth a resolution calling for all Southern Baptists to remove their children from public schools. Although rejected, the 2006 proposal stated that "children are our most important mission field, and the overwhelming majority of Christians have made the government school system their children's teacher," and "88 percent of the children raised in evangelical homes leave church at the age of 18, never to return."[20]

Contrast this with a recent survey of adults who were educated at home. About 94% strongly agreed or agreed with the statement "My religious beliefs are basically the same as those of my parents."[21]. What an awesome confirmation of the truth found in Proverbs 22:6 that if we train up a child in the way they should go, when they are old, they will not depart from it. Children brought up this way are much less likely to exchange God's truth for man's lies.

As Christian parents live in covenant with their children, their educational plan must be a plan of restoration[22], transforming children by the renewing of their minds (Romans 12:12) and training them to make every thought captive to the obedience of Christ (II Corinthians 10:5). We must teach them to be like David, a man after God's own heart (I Samuel 13:14). Christians must never forget that God created man in His image. Man is therefore an analogue of God, which is why it is so important for us to pattern our thoughts after God's thoughts.[23] Through obedience, man will become free, having the desire and ability to accomplish God's will.[24]

Critics may be thinking right now, "transforming minds, holding every thought captive; that's brainwashing!" Well, yes it is, but you can make the "brainwashing" argument against anyone who tries to educate another person. We are going to "wash" our children's brains with something, and I would rather see them washed with God's purifying message than with the eternally dirty rags of secularism, Darwinism and other false religions.

Conclusion

God created, man fell, Christ redeemed, and we are now in the period where "God's people are called to participate in a follow-up battle, pushing the enemy back and reclaiming territory for God."[25] If you believe this, then do not doubt the importance of a Christian education in the restoration of His kingdom on earth, as it is in Heaven (Matthew 6:10).

> A Christian education brings a message of hope, helping the student realize their calling as an active and creative participant in His awesome plan of restoration.

A proper Christian education begins on a personal foundation, expressing the personality of God, and allowing the student to realize they were created in His likeness. Because children are an analogue of God, they have the ability to be extremely creative, because the One they are patterned after is the best Creator of all[26]. A Christian education brings a message of hope, helping the student realize their calling as an active and creative participant in His awesome plan of restoration.

Christ's redemptive work paid the penalty for the sins of all who believe, saving us from eternal death. It is by His grace alone that we are saved (Ephesians 2:8-9), and it is His call as to whether we receive this gift (Acts 2:39). A faithful Christian also knows that faith, without works, is dead (James 2:17), and that he will reward each of us according to what we have done (Revelation 22:12). Although we know who wins the race, He calls us to run it with perseverance (Hebrews 12:1), and at times, to defend against evil (Ephesians 6:11). And it is through a Christian education that we teach our children to be faithful, to work joyfully, to be creative, to defend, to hope, and to persevere.

Final Summary

Well, now we have reached the end of the book. Let us conclude with a quotation from Francis Bacon, covered in Chapter 1:

> *A man cannot be too well studied in the book of God's word or in the book of God's works, divinity or philosophy, but rather endeavor an endless progress or proficience in both; only let men beware that they apply both to charity, and not to swelling, to use and not to ostentation; and again, that they do not unwisely mingle or confound these learnings together.[27]*

I Corinthians 13:2 says

> *And though I have the gift of prophecy, and understand all mysteries and all knowledge, and though I have all faith, so that I could remove mountains, but have not love, I am nothing.*

Don't just put the education you have received from this book up on a shelf. Use it, review it, and apply it. Teach others about the untruths of Darwin's words and the truths of God's words and works. If you are in a school that uses secular biology textbooks, don't hesitate to show your teacher and fellow classmates the errors it

> Teach others about the untruths of Darwin's words and the truths of God's words and works.

contains. Ask them questions and get them to think critically about what they are reading and believing. However, whatever you do, do it out of love, otherwise God considers you as nothing.

If you set out to show people about the untruths you have learned, especially people in a state-supported institution, you will likely be perceived as an enemy of the state and hated for your beliefs. However, don't fight hate with hate, but instead follow the wisdom of Solomon and *"If your enemy is hungry, give him bread to eat; and if he is thirsty, give him water to drink; For so you will heap coals of fire on his head, and the Lord will reward you"* (Proverbs 25:21-22). In other words, discuss these issues with him over a meal! God has a special plan for you, and if you follow Him fearlessly, serving Him and others in love, then God promises that your rewards will be many.

God has created us to serve and enjoy him forever. Through Scripture, prayer, and a Christian education, we are transformed by the renewing of our minds, and we will more readily discern when Truth is exchanged for a lie.

In the end, every knee will bow and every tongue confess that Jesus is Lord (Philippians 2:10-11), but until then, there are things He wants us to do. So what are you waiting for? Go.

NOTES

Introduction

1   Nickel, James. *Mathematics, Is God Silent?* 2nd ed. Vallecito, California: Ross House Books, 2001, p. 150.

2   Dunham, William. *Euler, The Master of Us All.* Mathematical Assoc. of America, 1999, p. xxvi.

3   Nickel, p. 151

4   Euler, Leonhard. *Letters to a German Princess*, Vol. 1. London: Thoemmes Press, 1997 reprint of the 1795 edition, p. 512.

Chapter 1 Deductive Reasoning

1   Nickel, James. *Mathematics, Is God Silent?* Vallecito, California: Ross House Books, 2001, p. 52.

2   Ibid, p. 53

3   Starr, Cecie and Ralph Taggart. *Biology, The Unity and Diversity of Life.* Belmont, California: Brooks/Cole-Thomson Learning, 2004, p. 15.

4   Gelb, Michael. *Discover Your Genius.* New York: HarperCollins Publishers, Inc, 2002, p. 145.

5   Linklater, Andro. *Measuring America.* New York: Walker Publishing Company, Inc, 2002, p. 26

6   Linklater, *Measuring America*, p. 18

7   Newton, Isaac. *The Principia, reprint of 1687 edition.* Amherst, New York: Prometheus Books, 1995, p. 440.

8   www.criticalthinking.org

9   Bacon, Francis. *Francis Bacon/edited by Brian Vickers.* Oxford: Oxford University Press, 2002, p. xlii.

10   Ibid, p. 126.

11   Pearcy, Nancy. *Total Truth.* Wheaton, Illinois: Crossway Books, 2004, p. 301.

12   Bacon, *Francis Bacon/edited by Brian Vickers*, p. 140.

13   Ibid, p. 144

14   Barlow, Nora. *The Autobiography of Charles Darwin.* New York: W.W. Norton & Company, 1958, p. 70.

[15] Ibid, p. 47.
[16] Ibid, p. 58.
[17] Ibid, p. 241.
[18] Ibid, p. 72.
[19] Ibid, p. 57.
[20] Ibid, p. 87.
[21] Darwin, Charles. *The Descent of Man, 2nd ed, 1874*. Amherst, NY: Prometheus Books, 1998, p. 62.
[22] Starr and Taggart. *Biology*, p. 3.
[23] Hedtke, Randall. *The Great Evolution Curriculum Hoax*. Phoenix, AZ: ACW Press, 2002, p. 42.
[24] Huckins, Kyle. 2003. Bias Causes Most to Bend, Break. *How do You Find a Christian College? Magazine by* www.findachristiancollege.org, p. 9.
[25] Nickel, James D. *Mathematics: Is God Silent?* Vallecito, CA: Ross House Books, 2001, p. 85.
[26] Smith, Ralph A. *Trinity and Reality*. Canon Press, Moscow, Idaho. 2004, p. 64.

Chapter 2 Inductive Reasoning and Critical Thinking

[1] Barlow, Nora. *The Autobiography of Charles Darwin*. New York: W.W. Norton & Company, 1958, p. 70.
[2] Bacon, Francis. *Francis Bacon/edited by Brian Vickers*. Oxford: Oxford University Press, 2002, p. 144.
[3] Ibid, p. 144.
[4] Barlow, *The Autobiography of Charles Darwin*, p. 159.
[5] Bacon, *Francis Bacon/edited by Brian Vickers*, p. 153.
[6] Ibid, p. 153.
[7] Dunham, William. *Euler, The Master of Us All*. USA: Mathematical Assoc. of America, 1999, p. xiii.
[8] Slack, Charles. *Noble Obsession*. New York: Hyperion Books, 2002.
[9] www.goodyear.com
[10] Hearne, Chester G. *Tracks in the Sea*. New York: International Marine/McGraw-Hill, 2002.
[11] homepages.rootsweb.com/~lpproots/Fountaine/mfm-01.htm
[12] Maury, Matthew Fontaine. *The Physical Geography of the Sea, 1855*. Mineola, NY: Dover Publications, 2003, p. 69.
[13] Bacon, *Francis Bacon/edited by Brian Vickers*, p. 126.
[14] Stoddard, Gloria May. *Snowflake Bentley*. Shelburne, VT: New England Press, 1985.
[15] The World Book Encyclopedia. Field Enterprises Educational Corporation, 1971, vol. 17.

[16] Bacon, *Francis Bacon/edited by Brian Vickers*, p. 153.

[17] Ibid, p. 141.

[18] Facione, Peter. 1998. Critical Thinking: What It Is and Why It Counts. *California Academic Press.*

[19] Ibid, p. 2.

[20] Bacon, *Francis Bacon/edited by Brian Vickers*, p. 147

[21] Ibid, p. 146.

[22] Ibid, p. 147.

[23] Discovery Institute. Center for Science and Culture. *A preliminary analysis of the treatment of evolution in biology textbooks currently being considered for adoption by the Texas State Board of Education.* Seattle: Discovery Institute, 2003, p. 49.

[24] Euler, Leonhard. *Letters of Euler to a German Princess, Vol. 1 & 2, 1795.* Great Britain: Thoemmes Press, 1997, vol. 1, p. 510.

Chapter 3 Fundamentals of DNA and Genetics

[1] Starr, Ralph and Cecie Taggart. *Biology, The Unity and Diversity of Life, 10th ed.* USA: Thomson Learning, 2004, p. 227.

[2] www.garlandscience.com

[3] Starr and Taggart, *Biology*, p. 221.

[4] Morris, Henry M. *Men of Science/Men of God* (El Cajon, CA: Master Books, 1988), p. 59.

[5] Starr and Taggart, *Biology*, p. 180.

[6] Ibid, p. 181.

[7] www.thewordfortoday.org

Chapter 4 Mutations and Probability

[1] Starr, Ralph and Cecie Taggart. *Biology, The Unity and Diversity of Life, 10th ed.* USA: Thomson Learning, 2004, p. 271.

[2] Ibid, p. 227.

[3] Ibid, p. 235.

[4] Ibid, p. 279.

[5] Russell, Peter J. Genetics, 2nd ed. Glenview, IL: Scott, Foresman and Co., 1990, p. 802.

[6] Starr and Taggart, *Biology*, p. 235.

[7] Ibid, p. 235.

[8] Russell, Peter J. *Genetics, 2nd ed.* Glenview, IL: Scott, Foresman and Co., 1990, p. 800.

[9] Campbell, Neil A., *Biology*. Redwood City, CA: Benjamin/Cummings Publishing Co., Inc., 1990, p. 457.

[10] Starr and Taggart, *Biology*, p. 10.

[11] Ibid, p. 10.

[12] Sproul, R.C. *Not a Chance*. Grand Rapids: Baker Books, 1994, p. 39.

[13] Hanegraaff, Hank. *The Face That Demonstrates The Farce of Evolution*. Nashville: W Publishing Group, 1998, p. 71.

[14] Ibid, p. 71.

[15] Starr and Taggart, *Biology*, p. 3.

[16] Darwin, Charles. *On the Origin of Species, 1st ed., 1859*. London: Harvard University Press, 1964, p. 281.

[17] Barlow, Nora. *The Autobiography of Charles Darwin*. New York: W.W. Norton & Company, 1958, p. 58.

[18] Bacon, Francis. *Francis Bacon/edited by Brian Vickers*. Oxford: Oxford University Press, 2002, p. 144.

[19] Ibid, p. 141.

[20] Johnson, Phillip E. *The Wedge of Truth*. Downer's Grove, IL: InterVarsity Press, 2000, p. 48.

Chapter 5 Microevolution

[1] Strachan, Tom and Andrew P. Read. *Human Molecular Genetics, 3rd ed*. New York: Garland Science, 2004, p. 463.

[2] Ibid, p. 462.

[3] Starr, Ralph and Cecie Taggart. *Biology, The Unity and Diversity of Life, 10th ed*. USA: Thomson Learning, 2004, glossary.

[4] Ibid, p. 279.

[5] Ibid, glossary.

[6] Barlow, Nora. *The Autobiography of Charles Darwin*. New York: W.W. Norton & Company, 1958, p. 58.

[7] Ibid, p. 47.

[8] Ibid, p. 109.

[9] Schreiber, Bernhard. *The Men Behind Hitler*. e-book: www.toolan.com/hitler/index.html.

[10] Ibid.

[11] Darwin, Charles. *On the Origin of Species, 1st ed., 1859*. London: Harvard University Press, 1964, p. 4-5.

[12] Coldwater Media. 2002. Icons of Evolution, Dismantling the Myths.

[13] Starr and Taggart, *Biology*, glossary.

[14] Ibid, glossary.

[15] Strachan and Read, *Human Molecular Genetics*, p. 470.

[16] Ibid, p. 470.

17 Discovery Institute. Center for Science and Culture, *A preliminary analysis of the treatment of evolution in biology textbooks currently being considered for adoption by the Texas State Board of Education.* Seattle: Discovery Institute, 2003, p. 16.

18 Starr and Taggart, *Biology,* p. 289.

19 Darwin, Charles. *The Descent of Man, 2nd ed., 1874.* Amherst, NY: Prometheus Books, 1998, p. 62.

20 Barlow, *The Autobiography of Charles Darwin,* p. 119.

21 Ibid, p. 119.

22 Ibid, p. 120.

23 Discovery Institute., *A preliminary analysis of the treatment of evolution in biology textbooks currently being considered for adoption by the Texas State Board of Education,* p. 16.

24 Starr and Taggart, *Biology,* p. 276.

Chapter 6 Macroevolution

1 Starr, Ralph and Cecie Taggart. *Biology, The Unity and Diversity of Life, 10th ed.* USA: Thomson Learning, 2004, p. 3.

2 Malthus, Thomas. *An Essay on the Principles of Population, 1798.* Amherst, New York: Prometheus Books, 1998, p.14.

3 Ibid, p. 20.

4 Schaefer, Henry F. III. Science and Christianity, Conflict or Coherence? Watkinsville, GA: Univ. of Georgia Printing, 2003, p. 105.

5 Starr and Taggart, *Biology,* p. 305.

6 Ibid, p. 294.

7 Ibid, p. 294.

8 Ibid, p. 294.

9 Ibid, glossary

10 Ibid, p. 306.

11 Ibid, p. 306.

12 Ibid, p. 306.

13 Ibid, p. 269.

14 Darwin, Charles. *On the Origin of Species, 1st ed., 1859.* London: Harvard University Press, 1964, p. 280.

15 Hanegraaff, Hank. *The Face That Demonstrates The Farce of Evolution.* Nashville: W Publishing Group, 1998, p. 33.

16 Starr and Taggart, *Biology,* p. 308.

17 Discovery Institute. Center for Science and Culture, *A preliminary analysis of the treatment of evolution in biology textbooks currently being considered for adoption by the Texas State Board of Education.* Seattle: Discovery Institute, 2003.

[18] Starr and Taggart, *Biology*, p. 310

[19] Vail, Tom. *Grand Canyon, a Different View*. Green Forest, AR: Master Books, 2003.

[20] Maury, Matthew Fontaine, *The Physical Geography of the Sea, 1855*. Mineola, NY: Dover Publications, p. 241.

[21] Starr and Taggart, *Biology*, p. 312..

[22] Ibid, p. 315.

[23] Discovery Institute. Center for Science and Culture, *A preliminary analysis of the treatment of evolution in biology textbooks currently being considered for adoption by the Texas State Board of Education*. Seattle: Discovery Institute, 2003, p. 48.

[24] Starr and Taggart, *Biology*, p. 315.

[25] Darwin, Charles. *The Descent of Man, 2nd ed., 1874*. Amherst, NY: Prometheus Books, 1998, p. 62.

[26] Barlow, Nora. *The Autobiography of Charles Darwin*. New York: W.W. Norton & Company, 1958, p. 240.

[27] Ibid, p. 87.

[28] Ibid, p. 58.

[29] Ibid, p. 57.

[30] Ibid, p. 87.

Chapter 7 Origins and the Age of the Earth

[1] Darwin, Francis. *The Life and Letters of Charles Darwin, Vol. 2*, p. 202.

[2] Starr, Ralph and Cecie Taggart. *Biology, The Unity and Diversity of Life, 10th ed.* USA: Thomson Learning, 2004, p. 327.

[3] Ibid, p. 327.

[4] Discovery Institute. Center for Science and Culture, *A preliminary analysis of the treatment of evolution in biology textbooks currently being considered for adoption by the Texas State Board of Education*. Seattle: Discovery Institute, 2003, p. 7.

[5] Starr and Taggart, *Biology*, p. 329.

[6] Discovery Institute, *A preliminary analysis of the treatment of evolution in biology textbooks currently being considered for adoption by the Texas State Board of Education*, p. 47.

[7] Ibid, p. 50.

[8] Ibid, p. 7.

[9] Morris, Henry M. *That Their Words May Be Used Against Them*. Green Forest, AR: Master Books, 1997.

[10] Starr and Taggart, *Biology*, p. 309.

[11] www.reasons.org

[12] Hummel, Charles E. *The Galileo Connection*. Downers Grove, IL: InterVarsity Press, 1986, back cover.

13 Ibid, p. 99.
14 www.icr.org
15 www.icr.org/pubs/imp/imp-353.htm
16 Arizona Highways Magazine, January 1993, pg. 10-11.
17 www.rbgsyd.gov.au
18 Starr and Taggart, *Biology*, p. 326.

Chapter 8 Biblical Creationism

1   Starr, Ralph and Cecie Taggart. Biology, The Unity and Diversity of Life, 10th ed. USA: Thomson Learning, 2004, p. 15.
2   Discovery Institute, Center for Science and Culture. A preliminary analysis of the treatment of evolution in biology textbooks currently being considered for adoption by the Texas State Board of Education. Seattle: Discovery Institute, 2003, p. 49.
3   Hummel, Charles E. The Galileo Connection. Downers Grove, IL: InterVarsity Press, 1986, p. 106.
4   Ibid, p. 105.
5   Darwin, Charles. The Descent of Man, 2nd ed., 1874. Amherst, NY: Prometheus Books, 1998, p. 62.
6   Morris, Henry M. The Defender's Study Bible, King James Version. Iowa Falls, IA: World Bible Publishers, 1995.
7   Schaefer, Henry F. III,. Science and Christianity, Conflict or Coherence? Watkinsville, GA: Univ. of Georgia Printing, 2003.
8   Crane, Nicholas. Mercator, the Man who Mapped the Planet. New York: Henry Holt and Company, LLC, 2003, p. 47.
9   Ibid, p. 44.
10 Ibid, p. 47.
11 Ibid, p. 93.

Chapter 9 Intelligent Design

1   Strachan, Tom and Andrew P. Read. *Human Molecular Genetics, 3rd ed.* New York: Garland Science, 2004, p. 468.
2   Ibid, p. 469.
3   Johnson, Phillip E.. *The Wedge of Truth.* Downer's Grove, IL: InterVarsity Press, 2000, p. 15.
4   www.arn.org
5   Barlow, Nora. *The Autobiography of Charles Darwin.* New York: W.W. Norton & Company, 1958, p. 59.
6   Ibid, p. 87.

7   Ibid, p. 87.

8   Behe, Michael. *Darwin's Black Box: The Biochemical Challenge to Evolution.* New York: Free Press, 1996.

9   Johnson, *The Wedge of Truth,* p. 48.

10   www.arn.org

11   Ibid

12   Hedtke, Randall. *The Great Evolution Curriculum Hoax.* Phoenix, AZ: ACW Press, 2002, p. 43.

13   Barlow, *The Autobiography of Charles Darwin,* p. 87.

14   Johnson, *The Wedge of Truth,* p. 176.

Chapter 10 Evolution and Scripture Cannot Both be True

1   Darwin, Charles. *The Descent of Man, 2nd ed.,* 1874. Amherst, NY: Prometheus Books, 1998, p. 62.

2   Darwin, Charles. *On the Origin of Species, 1st ed., 1859.* London: Harvard University Press, 1964, p. 490.

3   Barlow, Nora. *The Autobiography of Charles Darwin.* New York: W.W. Norton & Company, 1958, p. 87.

4   Darwin, *The Descent of Man,* p. 641.

Chapter 11 The Effects of an Evolutionary Worldview on Society

1   Darwin, Charles. *The Descent of Man, 2nd ed, 1874.* Amherst, NY: Prometheus Books, 1998, p. 62.

2   Schreiber, Bernhard. *The Men Behind Hitler.* Internet book: www.toolan.com/hitler/index.html, date unknown, Chapter 1.

3   Darwin, *The Descent of Man,* p. 642.

4   Schreiber, *The Men Behind Hitler,* Chapter 1.

5   Salak, Kira. *Places of Darkness.* National Geographic Adventure Magazine, Dec. 2003/Jan 2004, p. 88.

6   Ibid, p. 89.

7   Mthembu-Salter, Gregory. *Rwanda,* www.selfdetermine.org/conflicts/rwanda_body.html

8   Sehene, Benjamin. *The Ethnic Trap.* Internet book: http://victorian.fortunecity.com/cloisters/870/theethnictrap.html, date unknown.

9   Ibid

10   Schreiber, *The Men Behind Hitler,* Chapter 3.

11 Hitler, Adolf, *Mein Kampf.* Internet book: www.stormfront.org/books/mein_kampf, vol. 1, Ch. 11.

12 Ibid

13 Dew, Diane. Black *Genocide: Planned Parenthood's Evil Roots,* http://dianedew.com/black.htm

14 Ibid

15 National Right to Life. www.nrlc.org/abortion/facts/abortionstats.html

16 Grant, George, *Killer Angel.* New York: Ars Vitae Press, 1995, p. 3.

17 Ibid, p. 3.

18 Ibid, p. 3.

19 Ibid, p. 70.

20 Ibid, p. 71.

21 Ibid, p. 72.

22 Ibid, p. 72.

23 Linder, Doug. "State v. John Scopes. "The Monkey Trial," www.law.umkc.edu/faculty/projects/ftrials/scopes/evolut.htm

24 Johnson, Phillip E. *The Demise of Naturalism.* World Magazine, April 3, 2004, p. 36.

25 Hedtke, Randall,. *The Great Evolution Curriculum Hoax.* Phoenix, AZ: ACW Press, 2002, p. 42.

26 Scott, Darrell, *Oct. 21, 2003 letter to Texas State Board of Education,* http://www.strengthsandweaknesses.org/PDFs/D.Scott.Oct.13.PR.2.pdf

Chapter 12 Education

1 http://elbourne.org/

2 Pearcey, Nancy. *Total Truth.* Wheaton, Illinois: Crossway Books, 2004, p. 95.

3 Nickel, James D. *Mathematics: Is God Silent?* Vallecito, CA: Ross House Books, 2001, p. 85.

4 Pearcey, *Total Truth*, p. 155.

5 Euler, Leonhard. *Letters of Euler to a German Princess, Vol. 1 & 2, 1795.* Great Britain: Thoemmes Press, 1997, vol. 1, p. 187

6 Bacon, Francis. *Francis Bacon/edited by Brian Vickers.* Oxford: Oxford University Press, 2002, p. 140.

7 Phillips, Melanie. *All Must Have Prizes.* London: Little, Brown and Company, 1996, p. 197.

8 Dewey, John. *Democracy and Education.* Online book: http://www.ilt.columbia.edu/publications/dewey.html, 1916, Chapter 1.

9 Phillips, *All Must Have Prizes*, p. 210-211.

10 Ibid, p. 212.

11 Klicka, Christopher J. The Right Choice, Homeschooling. Gresham, Oregon: Noble Publishing Associates, 1992, p. 116.
12 Van Til, Cornelius. *Essays on Christian Education*. Phillipsburg, NJ: Presbyterian and Reformed Publishing Company, 1979, p. 131.
13 Ibid, p. 133.
14 Ibid, p. 139.
15 Ibid, p. 198.
16 Ibid, p. 199.
17 Ibid, p. 204.
18 Ibid, p. 186.
19 www.sbc.net/resolutions/amResolution.asp?ID=314
20 Resolution on developing an exit strategy from the public schools that would give particular attention to the needs of orphans, single parents, and the disadvantaged. Submitted to the Southern Baptist Convention, April 24, 2006. www.exodusman-date.org/20060425-sbc-resolution/2006-resolution.doc
21 www.nheri.org
22 Van Til, *Essays on Christian Education*, p. 143.
23 Ibid, p. 150.
24 Ibid, p. 155.
25 Pearcey, *Total Truth*, p. 91.
26 Ibid, p. 150.
27 Bacon, *Francis Bacon/edited by Brian Vickers*, p. 126.

ABOUT THE AUTHOR

David Shormann holds a B.S. in Aerospace Engineering and a M.S. in Marine Chemistry from The University of Texas, and a PhD in Aquatic Science from Texas A&M University. In 1997 he and his wife Karen founded Genesis Science, Inc., a company that provides math and science classes to home-educated students. In 2001 they founded DIVE, LLC, a company providing Christian-based math and science lectures on CD-ROM. Dr. Shormann has taught extensively on the creation/evolution controversy. He currently lives with his family near Houston, TX, and serves as a deacon for St. David's Church.

For a free study guide and other information, please visit www.exchangeoftruth.com

INDEX

Note: Italicized f following page locators indicate figures.

biblical framework, reconciliation with, 89–90
theory of natural selection, 11
theory of segregation, 36
theory of uniformity, 68
thorium halos, 83
transfer RNA, 30
transposons, 41
truth
  classes of, xii
  and faith-based beliefs, 74–75
Tutsi peoples, 114–115

university system, 12
uranium halos, 83
Urey, Harold, 79

Van Til, Cornelius, 126, 127
vulcanization, of rubber, 17

wages of sin, 106–107
Wedge of Truth, The (Johnson), 49
Wilder-Smith, A.E., 37–38
wisdom, 3

978-0-595-42177-0
0-595-42177-6